KEPLER'S PHILOSOPHY AND THE NEW ASTRONOMY

Rhonda Martens

KEPLER'S PHILOSOPHY AND THE NEW ASTRONOMY

PRINCETON UNIVERSITY PRESS PRINCETON AND OXFORD

Published by Princeton University Press, 41 William Street
Princeton, New Jersey 08540
In the United Kingdom: Princeton University Press
3 Market Place, Woodstock, Oxfordshire OX20 1SY

Library of Congress Cataloging-in-Publication Data

Martens, Rhonda, 1964–
Kepler's philosophy and the new astronomy / Rhonda Martens.
p. cm.
Includes bibliographical references and index.
ISBN 0-691-05069-4 (alk. paper)
1. Kepler's laws. 2. Astronomy. I. Title.
QB355.3 .M37 2000
521'.3—dc21 00-036681

This book has been composed in Janson

The paper used in this publication meets the minimum
requirements of ANSI/NISO Z39.48-1992 (R 1997)
(*Permanence of Paper*)

www.pup.princeton.edu

Printed in the United States of America

10 9 8 7 6 5 4 3 2 1

For Hedy and John Martens, and for Jason Holt

Contents

Figures and Tables

Figures

Tables

Acknowledgments

THIS BOOK began as an attempt to understand how it could be that the person who wrote the *Mysterium cosmographicum* could also be the same who wrote the *Astronomia nova.* John Thorp and William Harper pointed me in the direction of this problem, which became the primary theme of my doctoral dissertation at the University of Western Ontario. William Harper, my dissertation supervisor, guided my efforts with grace and seemingly infinite patience. William's constructive criticisms, emotional support, and generosity with his time, were the sine qua non of this book, and I am lucky to have worked with him. I am grateful to John Thorp for his tutelage in matters of ancient Greek music theory and Latin, and his support of my project. Howard Plotkin offered an excellent critique of my dissertation, and chapter 4 in particular benefited enormously from his efforts. Owen Gingerich was the external examiner for my dissertation, and I am thankful for his support and direction.

I addressed the business of turning my thesis into a book while on an Postdoctoral Fellowship at the University of Chicago. I owe a tremendous debt of gratitude to Noel Swerdlow, whose patronage and advice have been invaluable. Noel was extremely giving not only of his expertise in the history of astronomy, but also his time and marvelous sense of humour. Writing a book taxes both time and spirit, and Noel's support made the latter a renewable resource. Noel also generously offered the wonderful diagrams shown in figures 1.2–1.9, 2.1, 3.1, 3.2, 4.1–4.3, 4.5, 4.7–4.10, 7.1–7.3.

Various sections of this book were read at conferences and colloquia and discussed with friends, and I have received valuable feedback. In particular, I thank Nicholas Jardine, Howard Stein, Dan Garber, David Millar, Elaine Landry, Eric Schliesser, Erik Curiel, Steven Secker, Tim Kenyon, Bryce Bennett, William Vanderburgh, and Fransisco Flores. James Voelkel has been very generous with the results of his excellent study of the Kepler-Fabricius correspondence.

Last, but definitely not least, I owe much to Jason Holt. Not only has he been extremely supportive of this project from start to finish, but he spent countless hours poring over various drafts of this book and copyediting the penultimate version. Any elegant turns of phrase are his doing, and the final product has been improved immeasurably by his efforts. I dedicate this book to him and to my parents, John and Hedy Martens.

The discussion in chapter 4 on Kepler's libration theory and his "beautiful demonstration" is reprinted [nearly verbatim] from "Kepler's Solution to the Problem of a Realist Celestial Mechanics," *Studies in History and Philosophy of Science* 30 (1999): 384–85, with permission from Elsevier Science.

An earlier version of chapter 5 was published as "Kelper's Use of Archetypes in his Defense against Aristotelian Scepticism," in *Mysterium Cosmographicum, 1596–1996*, edited by J. Folta, Acta Historiae Rerum Naturalium Necnon Technicarum: Prague Studies in the History of Science and Technology, vol. 2 (Prague: National Technical in Prague, 1998), pp. 74–89.

Financial support from this project was provided by the Social Sciences and Humanities Research Council of Canada for both my doctoral and postdoctoral work.

Abbreviations of Works Frequently Cited

AN Kepler, J. ([1609] 1992). *New Astronomy.* Translated by W. H. Donahue. New York: Cambridge University Press.

Apologia Kepler, J. ([1858] 1984). *Apologia pro Tychone contra Ursum.* Translated by N. Jardine. In *The Birth of History and Philosophy of Science: Kepler's "A Defense of Tycho against Ursus" with Essays on Its Provenance and Significance.* Cambridge: Cambridge University Press.

Epitome Kepler, J. ([1618–21] 1952). *Epitome of Copernican Astronomy: Books IV and V.* Translated by C. G. Wallis. In *Great Books of the Western World,* vol 16, edited by Robert Maynard Hutchins, pp. 839–1004. Chicago: Encyclopaedia Britannica.

HM Kepler, J. ([1619] 1997). *The Harmony of the World.* Translated with an introduction and notes by E. J. Aiton, A. M. Duncan, and J. V. Field. Philadelphia: American Philosophical Society.

KGW Kepler, J. (1937–). *Johannes Kepler Gesammelte Werke.* Edited by V. W. Dyck, M. Caspar, and F. Hammer. Munich: Beck.

MC Kepler, J. ([1596] 1981). *The Secret of the Universe.* Translated by A. M. Duncan with notes by E. J. Aiton. New York: Abaris Books.

KEPLER'S PHILOSOPHY AND THE NEW ASTRONOMY

Introduction

One of the most fertile minds ever, Johannes Kepler (1571–1630) made valuable contributions to every field he addressed. He changed the face of astronomy by abandoning principles that had been in place for some 2,000 years, made important discoveries in optics and mathematics, and even constructed astrological charts renowned for their uncanny accuracy. In addition, he was an uncommonly good philosopher.

One tends not to hear much about Kepler's philosophical prowess, probably because he did not actually write a treatise specifically on philosophy; his philosophical views were usually advanced as solutions to problems in other disciplines (the *Apologia pro Tychone contra Ursum* is an exception, but its content was constrained by Tycho Brahe). Consequently, in order to encounter Kepler's philosophical views, one has to study his treatment of other subjects. Kepler's astronomical works provide a particularly fruitful source for his epistemology and methodology of natural philosophy, what we now call "science."

For one thing, Kepler became a Copernican at a time when there was no empirical evidence that would support the Copernican over the well-entrenched Ptolemaic system (and, later, over the Tychonic system). His initial reasons for preferring Copernicanism, therefore, were extraempirical and were drawn in particular from metaphysics and methodology. Consequently, in his early astronomical works one finds interesting arguments on the logic of drawing predictions from theory, the virtue of the explanatory power of a theory, what makes a theory testable, and so on. Later, he added arguments from physics to his arsenal, which raised other philosophical issues.

For another thing, astronomers were well aware of certain philosophical problems having to do with the making and interpretation of observations. It was standard practice to *process* observations in the computation of planetary positions. One had to calculate the effects of refraction and to consider whether the body was observed at the horizon or directly overhead (consider the familiar phenomenon of the change in size of the moon as it approaches the horizon). One could not always make the desired observation at precisely the right time—for example, if it was clouded over. One had to take into account that, because the object stood at such a distance from the observer, a small error in observation could significantly influence the calculated position of the planet; and, if one was a Copernican, then the position of the Earth had to be taken into account. These and other considerations made it clear to astronomers that the relationship between observation and theory was quite complex and that a certain level of error was acceptable; even if an observation of a planet's position was five minutes off from the predicted position, this prediction could still be considered successful. In the *Astronomia nova* Kepler used discrepancies between prediction and observation that would have previously been within the acceptable margin of error to revolutionize astronomy. In order to do this, he needed to address the issue of how to make observations yield more precise information.

For the Copernican astronomer, moreover, the issue of the interpretation of observations was particularly keen. After all, it *looks* like the sun goes around the Earth. Yet being a Copernican does not mean that one ignores observations; rather, the question is how to determine a reliable and veridical method of interpreting observations, a concern that occupied Kepler throughout his career. Kepler's revolutionary approach to astronomy raised questions not only about the relationship between observation and theory but also about the very nature of astronomy and physics. Kepler argued that astronomy would progress only if it was grounded in physics. This position was unpopular at the time, prompting Kepler to address the philosophical problem of the relationships between disciplines.

In addition, the celestial physics Kepler developed was a mixture of Aristotelian qualitative physics and Archimedean mathematical mechanics. Because the relationship between mathematics and physics was an open question at the time (in contrast to today when it is taken for granted that one uses mathematics in physics), Kepler was subverting the standard methodology of physics, a situation that encouraged philosophical thought on the nature of physics. Aristotle had warned against the use of mathematics in physics. For Aristotle, the proper way to explain the behavior of physical objects was to deduce what they will do from a consideration of their essences. The mathematical features of objects are, for Aristotle, accidental rather than essential properties. Kepler's solution was to develop a metaphysics under which the mathematical properties of an object *are* its essential properties. One sees this position echoed in Descartes, who argued that the essential property of objects is their extension, which paved the way for a mechanical physics. His effort was sufficiently persuasive that Spinoza took it for granted that physics was the study of extension. This position became widespread enough that Leibniz felt the need to argue that physics required nonmathematical as well as mathematical principles.

Kepler's mathematical metaphysics is the central theme of this book, because it is this metaphysics that provided the inspiration and justification for the methodological innovations required by the new astronomy. Kepler believed that the world was created by God to express divine aesthetics, what Kepler called "Archetypes." This aesthetics was essentially geometrical in nature. If one knows the details of God's aesthetics, then one can infer a priori certain structural features of the world—for example, whether the world is Copernican or Ptolemaic. (The commitment to a priori knowledge about the world is a key feature of the early modern rationalist movement.) Conversely, if one knows the structural features of the world, then one can gain access to God's aesthetic plan.

In order to carry out this study, we need to look at Kepler's astronomy in some detail, which is interesting in its own right. Although no background in astronomy is presupposed, an assessment of some technical material is necessary if we are to get a sense of how Kepler applied the methodological principles developed from his mathematical metaphysics, and also of the problems that motivated the development of these principles. I have struggled to make Kepler's work as accessible as possible, a task most difficult with the material in chapter 6. Here I isolate commentary on technical details, which the reader can skip if desired.

Kepler's manuscripts have been collected in the *Johannes Kepler Gesammelte Werke* and *Joannis Kepleri astronomi opera omnia*. The *Mysterium cosmographicum*, *Apologia pro Tychone contra Ursum*, *Astronomia nova*, *Harmonice mundi*, selections of *Epitome astronomiae Copernicanae*, and selections of Kepler's correspondences have been translated into English. I have included citations to both the English translations (where available) and to the Latin and German in *Johannes Kepler Gesammelte Werke*.

The audience I had in mind when writing this text includes both historians and philosophers of science, metaphysics, and epistemology, and anyone interested in understanding the scientific revolution. One thing that becomes clear in this study is that the process of constructing a new discipline (in this case physical astronomy) brings to the forefront epistemological, methodological, and metaphysical issues, and I offer a detailed examination of this process. In this case study, we see in detail the complex interplay between data and theory, and the role theory plays in "turning data into evidence"[1] for that theory. We also see the role that metaphysics played for Kepler in justifying his epistemology as truth tracking. In addition, in Kepler's work we see the first and perhaps the most rigorous attempt to test empirically a metaphysical worldview, as well as one of the most explicit uses of metaphysics to further an empirical discipline. Thus, although Kepler's metaphysics did not start a new metaphysical tradition, it was instrumental in starting the tradition of physical astronomy, and is of historical and philosophical significance for that reason alone. Kepler is clearly a key figure in the scientific revolution, and, given that he explicitly stated that his astronomy was related to his metaphysics, this study should aid in understanding the manner and extent to which this is so.

Until recently one was likely to encounter one of two Keplers: Kepler the consummate scientist as depicted by Small ([1804] 1963), Dreyer ([1906] 1953), and Strong (1966); or Kepler the demented dream architect, the sleepwalker depicted most notably and popularly by Koestler ([1959] 1968). The Kepler one encounters often depends on whether an author focuses on or away from Kepler's "unscientific" interests. There are notable exceptions. Cassirer ([1927] 1963, 164–65), for example, takes note of Kepler's interest in archetypes and astrology without concluding that his method was unscientific; indeed, he proposed that Kepler's aesthetic conception of the world was instrumental in overcoming the mystical approach to nature. Recently there has been a resurgence of Kepler scholarship, and a number of excellent works have been published on him. This text is heavily indebted to these works, most notably those of Stephenson 1987; 1994; Jardine 1979; 1984; Gingerich 1973; 1975a; 1992; Gingerich and Voelkel 1998; Voelkel 1994; Field 1988; Caspar [1959] 1993; Harper and Smith 1995; Westman 1972; 1975; 1980; 1984; and Mittelstrass 1972. Current Kepler scholarship tends to acknowledge that there is a relationship between Kepler's scientific and "unscientific" interests, and studies on his metaphysics (e.g., Lindberg 1986; Field 1988; Stephenson 1994; and Kozhamthadam 1994), and his astrology (e.g., Rosen 1984) have appeared. In addition, several texts have explored his astrology and archetypal cosmology in relation to his astronomy (e.g., Pauli 1955; Holton 1956; Koyré [1961] 1973; Simon (1975) and have indicated the importance of these interests in Kepler's conversion to Copernicanism. Holton, in particular, emphasizes the way that Kepler appealed to his archetypes when his physical arguments failed. In this text I approach Kepler from a somewhat different perspective; I am interested in those aspects of the relationship between Kepler's metaphysics, epistemology, and physical astronomy which promise the greatest philosophical yield. In particular, I examine Kepler's use of metaphysics to resolve the epistemological and methodological problems in astronomy generally acknowledged in his time.

In addition to complementing the recent Kepler scholarship, this text is intended to add detail and nuance to the recent broad discussions of the scientific revolution. On the whole I have found such narratives like John Henry's (1997) and Steven Shapin's (1996) to be quite responsive to recent detailed studies of key figures in their cultural contexts, some of which exhibit their so-called unscientific interests. The impression one often gets from these overviews featuring such unscientific aspects, however, is that the scientific revolution was a far less rationally motivated transition than had previously been supposed. My aim is to show that, in Kepler's case, such unscientific influences were used ingeniously to resolve the conceptual difficulties that faced his scientific program. While it is a welcome trend that recent narratives pay attention to the writings of Kepler erstwhile viewed as unscientific and therefore irrelevant, the Kepler one encounters in this book, which focuses on these "unscientific" ideas, is far more motivated by rational and scientific concerns than recent narratives would lead one to expect.

The book's first chapter is intended to provide just enough background information on Kepler's life and times to set the stage for the remaining discussion. After opening with a brief biography of Kepler (several detailed biographies of Kepler are available, most notably Caspar ([1959] 1993), I turn to the astronomy Kepler inherited. Here I review the technical apparatus available to Kepler, as well as prominent views at that time on the status and interpretation of astronomy, especially in light of the Copernican and Brahean challenges to traditional Ptolemaic astronomy. The chapter concludes with a discussion of the philosophical trends that most likely influenced Kepler's thought.

The remaining chapters focus on Kepler's astronomical works as well as those on metaphysics and epistemology most relevant to his astronomy. Because my interest is also in the evolution of Kepler's metaphysical and epistemological thought, especially in how it changed in response to new astronomical discoveries, Kepler's works will be discussed in the order in which they were written.

Chapter 2 discusses Kepler's *Mysterium cosmographicum*, his first book, which was written before he developed proficiency as an astronomer. The primary thesis of the *Mysterium* is that the number and relative sizes of planetary orbits can be explained by the hypothesis that the five regular solids (the cube, tetrahedron, dodecahedron, icosahedron, octahedron), if nested inside each other in an order determined by their importance, roughly correspond to the relative sizes of the planetary orbits. This may seem a peculiar hypothesis, but as we shall see, Kepler's archetypal vision not only supported the Copernican hypothesis but underwrote his approach to natural philosophy. In the *Mysterium* Kepler explicitly argued that an astronomical hypothesis required converging support from an archetypal model, and that such a model was available to the Copernican but not the Ptolemaic hypothesis. He argued for the superiority of the Copernican account on other grounds as well (e.g., fruitfulness), and I show how Kepler justified these grounds by appeal to his archetypal cosmology.

Chapter 3 is a brief discussion of the *Apologia*, in which Kepler argued against the skeptics that astronomical hypotheses, when properly constructed, interpreted, and tested, could provide us with a true account of planetary motion. The main problem facing the realist in astronomy was the availability of competing empirically adequate models. I argue that Kepler's answer to the skeptic, while appearing quite modern, was motivated by, and grounded in, his cosmology.

The *Mysterium* and the *Apologia*, however, are early works of Kepler, written before he attained the prowess as an astronomer for which he is famous. Because there is little mention of the archetypes in his mature work, the *Astronomia*, it would not be surprising if his youthful commitment to an archetypal world order had been abandoned. There are, nonetheless, two strong sources of evidence to the contrary; Kepler's later works, the *Harmonice mundi* and the *Epitome astronomiae Copernicanae* make frequent reference to the archetypes. Given that Kepler did not abandon his youthful vision of the universe, what remains to be explored is whether this vision played a significant part in his "war on Mars."[2] In chapter 4 I argue that it does, and not only to the extent that Kepler applied the method developed in the *Mysterium* and the *Apologia*. Kepler also used his archetypal cosmology to rule out preempirically various hypotheses—to determine which kinds of hypotheses were plausible. This approach permitted him to focus attention on avenues of inquiry that, in the end, turned out to be quite fruitful.

Chapter 5 is a reflection on the conceptual apparatus Kepler made use of in constructing his new physical astronomy and, as such, breaks the flow of the narrative to some extent. This reflection shows in detail the relationship between his three diverse early books, the *Mysterium*, the *Apologia*, and the *Astronomia*. I argue that Kepler's new astronomy involved methodological principles that were considered unsound during his time. I also argue that Kepler's archetypal cosmology enabled him to blend Platonic and Aristotelian intuitions, providing him with rhetorical means to defend his account from expected criticisms.

By the end of chapter 5, I hope to have made clear that Kepler's archetypal interests should not be viewed as separate from his interests in natural philosophy. They had an influence on his new physical astronomy at various levels, from the methodological to the rhetorical. By the time Kepler finished the *Astronomia*, however, he was aware that his polyhedral hypothesis was incomplete: it failed to explain the nonuniform elliptical motion of the planets. Kepler tried to rectify this problem in the *Harmonice*. Kepler's new archetypal model, however, was not fully compatible with his original cosmological commitments, and it became an open question whether the material world corresponded to the archetypal *precisely*. Given the importance of making this match, as outlined in the foregoing chapters, one wonders whether the *Harmonice* really was Kepler's "mind's favorite child" (Caspar [1959] 1993, 288). While the *Harmonice* is a remarkable book, the universe that Kepler believed he discovered was not the perfect one he had expected. In chapter 6 I trace Kepler's failed attempt to provide an archetypal justification of his new astronomy and the methodological implications of this failure.

Chapter 7 is a discussion of Kepler's *Epitome*, which is a significant work for three reasons. First, it was written as a textbook for a general audience and, as a result, is a good resource for determining what Kepler believed he needed to do to render his new astronomy plausible. Here we see Kepler's metaphysics featured prominently, which suggests that he considered it part of his rhetorical arsenal. Second, his mature physics, metaphysics, and astronomy are finally presented together in one work, which allows for further exploration not only of the evolution of his thought but also on how he conceived of the relationship between these diverse elements. Finally, since books IV through VII were written after the *Harmonice*, they give evidence of the tension between Kepler's physics, metaphysics, and astronomy that, I argue, was brought about by Kepler's difficulty in developing a complete archetypal account in the *Harmonice*.

The final chapter is a short commentary on the implications that the shortcomings of Kepler's approach to natural philosophy had for his interpretation of his astronomy as true. I conclude by briefly addressing the question of Kepler's influence on subsequent philosophical thought. Granted, his approach as a whole suffered difficulties, but a number of aspects of his philosophical thought had proved very fruitful. In particular I was interested in the manner and extent to which Descartes and Leibniz—who both shared much in common with Kepler's metaphysical thought, and who both were well aware of Kepler's works—commented on Kepler's philosophical positions.

This study makes clear that Kepler has much to offer philosophers and historians of science, epistemology, and metaphysics. Although Kepler did not write a philosophical treatise, his thought on these matters is particularly valuable precisely because they were developed in the context of providing solutions to significant problems facing an emerging science. A peculiar philosophy thus served to establish a lasting astronomy.

1

Kepler's Life and Times

Select Biography

Kepler's life was marked by personal misfortune, social instability, and intellectual fertility. He was hounded from all sides by physical illness, political and religious strife, and financial distress. An unrepentant Lutheran, he was threatened by the machinery of the Counter-Reformation; at the same time, he was an unorthodox Lutheran, and so was eventually excluded from Communion in the Lutheran Church. Denied a religious home, Kepler craved a politically stable environment in which to carry out his work, but a refuge was denied to him by the frequent wars that seemed to surround him. In 1621 he lamented this situation and its effect on astronomy:

> [T]imorous and unwarlike astronomy is warned by the conditions, dangers, terrors, disasters, and troubles to look round for assistance. She crossed in the year 1600 from Styria into Bohemia, so that just as she had put her first roots under the shelter of the Austrian house she might also grow to maturity under it. After being tossed to and fro there by the tempests of both civil and foreign wars . . . she returned to Austria. Would that she could have been honored there with the devoted attention of eminent minds (no less than by myself, who restored her) as much as she was accepted and favored with goodwill. Yet, alas, of what great goods do miserable mortals despoil one another, by their shameful itching for quarrels. (*MC*, 43; *KGW* VIII, 11)

Not all of the forces at work during Kepler's life were negative; the period in which he lived was also marked by lively intellectual activity. Kepler himself keenly felt the excitement of the times:

> After the birth of printing books became widespread. Hence everyone throughout Europe devoted himself to the study of literature. Hence many universities came into existence, and at once many learned men appeared so that the authority of those who clung to barbarism declined. . . . What shall I say of today's mechanical arts, countless in number and incomprehensible in subtlety? Do we not today bring to light by the art of printing every one of the extant ancient authors? . . . Through them there has today been created a new theology and a new jurisprudence; the Paracelsians have created medicine anew and the Coper-

> nicans have created astronomy anew. I really believe that at last the world is alive, indeed seething, and that the stimuli of these remarkable conjunctions did not act in vain. (Translated in Jardine 1984, 277–78; from *KGW* I, 330–32)

These were fertile times; indeed, one could probably not pick a more auspicious time for one with such astronomical gifts to have lived. He was born twenty-eight years after Nicolaus Copernicus's death (1473–1543) and had the opportunity to meet and work with the Danish astronomer Tycho Brahe (1546–1601), perhaps the greatest naked-eye observer in history. He was a contemporary of Galileo Galilei (1564–1642) and lived to hear of the invention of the telescope and to receive reports from Galileo of what was seen through it—the four moons of Jupiter, for example. Kepler himself had opportunity to use the telescope, and he established the reliability and veracity of telescopic observations through his study of the principles of how light passes through lenses (presented in a work called *Dioptrice*, 1611). As a result of the empirical work of Galileo and the theoretical work of Kepler, the advent of the telescope heralded a new era in the kinds of observations that could be made of the heavens. This age witnessed great advances in astronomy.

Kepler was born on December 27, 1571, to Heinrich and Katherina (née Gulderman) in Weil (today called Weil der Stadt). The picture of Kepler's family that Caspar drew from Kepler's account of the astrological causes of the personality traits of his family members is not particularly flattering. One gets the impression that Kepler's parents and grandparents, with whom Kepler lived as a small child, were harsh and uncaring (Caspar [1959] 1993, 34–36). Caspar ([1959] 1993, 36) speculates that this may have been the cause of Kepler's strong faith in God, to which he turned for solace. Kepler's piety, coupled with his keen intelligence, made him an ideal candidate for the clergy. While such a career choice brought prestige to the family, it also removed a heavy financial burden on the family; the seminary at Tübingen was "subsidized,"[1] so Kepler's schooling cost relatively little. As a result, he was able to pursue this path.

In 1589 Kepler entered Tübingen University. Tübingen was a Lutheran institution, and as such its curriculum was influenced by Martin Luther and Philip Melanchthon. Luther's thought seems less relevant to Kepler's development than that of Melanchthon, since Luther tended to speak of theology, natural philosophy, and philosophy as having distinct methods and subject matters (Methuen 1998, 62–68). By contrast, Melanchthon, like many of Kepler's teachers, saw nature as significantly revelatory. One could come to know about God by studying the created world (Methuen 1998, 69–99). Thus training in philosophy and natural philosophy was an integral part of theological training. Moreover, the study of mathematics was particularly important for Melanchthon due to his sympathy for the

Pythagorean and Platonic view that the ultimate nature of reality is mathematical. The tractability of the natural world to mathematical analysis is an indication of the extent to which it reflects the mind of God. The mathematical abilities of the human mind, according to Melanchthon, are also a reflection of God, allowing us to access God through the study of nature (Methuen 1998, 83). Consequently, mathematics was given a much higher status in Melanchthon's thought than was typical for the time.

Melanchthon's influence on the curriculum at Tübingen was substantial (Methuen 1998, 99–106), and so Kepler studied mathematics, astronomy, and physics, as well as ethics, dialectics, rhetoric, Greek, and Hebrew for two years before beginning studies in theology (Caspar [1959] 1993, 42). While the standard works of Aristotle were studied—Kepler reported that he became familiar with Aristotle's *Posterior Analytics* (a treatise on the proper method of natural philosophy) and *Physics* (*KGW* XIX, 328–29) during this time—Kepler also learned something of the Pythagoreans, Plato, and the Neoplatonists, although Methuen (1998, 221) argues that he likely learned less about these schools from the professors at Tübingen than is generally thought.

At Tübingen Kepler met perhaps the most important influence on his life, Magister Michael Maestlin, who taught mathematics and astronomy and was an accomplished and well-known astronomer himself.[2] Maestlin was also a Copernican, who inspired Kepler by speaking of the mathematical advantages the Copernican system had over the Ptolemaic (*MC*, 63; *KGW* I, 9). Thus began a mentorship and friendship that would last for many years.

Kepler's theological studies were interrupted by an invitation to teach mathematics at a Protestant seminary in Graz, an offer he was loath to take up but felt he could not refuse (*AN*, 183–84; *KGW* III, 108). This "interruption" turned out to be a blessing in disguise, for it was in Graz that he hit upon the idea that would be the central thesis in his first book, the *Mysterium*.

As he reported in the preface to the *Mysterium*, while teaching at Graz Kepler became obsessed with the question of why God had created exactly six planets, with the distances between them as they were (he had in mind the Copernican distances). He initially tried to find some rule of order in the numerical progression of the distances, but without success. Then he decided—a decision that would guide the course of his work—that God was a geometer rather than a numerologist. God would create the world to represent aesthetically interesting geometrical shapes. This came to him as a flash of inspiration while teaching a class:

> Therefore on the 9/19th of July in the year 1595 when I was going to show my audience the leaps of the great conjunctions [of Saturn and Jupiter] through

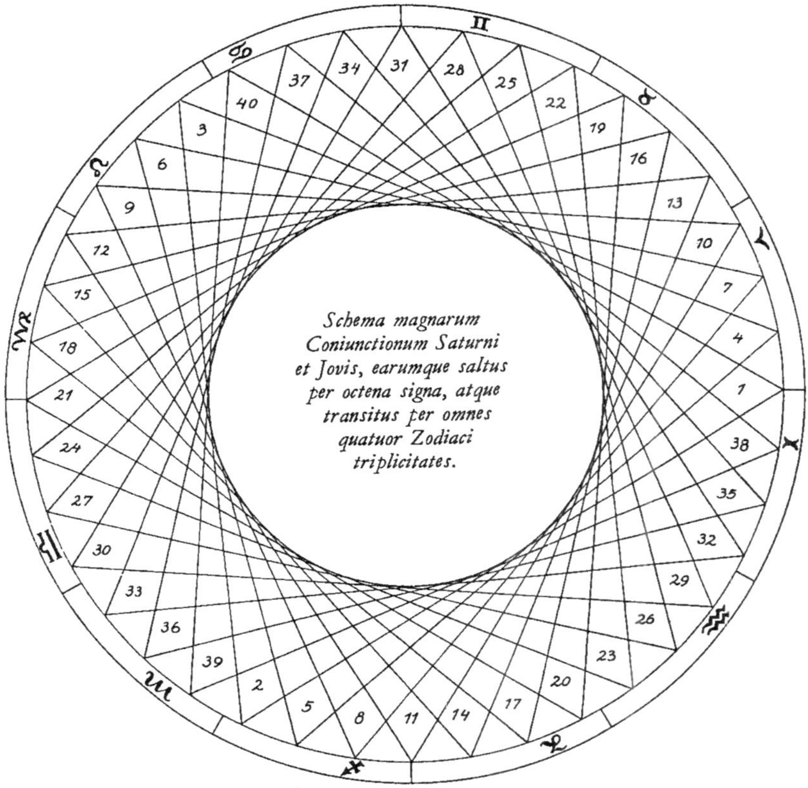

Figure 1.1. Kepler's Inscribed Triangles

> eight signs at a time, and how they cross step by step from one triangle to another, I inscribed many triangles, or quasi-triangles, in the same circle, so that the end of one was the beginning of another. Hence the points at which the sides of the triangles intersected each other sketched out a smaller circle. (*MC*, 65; *KGW* I, 11; figure 1.1 from *KGW* VIII, 26)

Kepler noticed that the ratio between the two circles was close to the ratio between the orbits of Saturn and Jupiter, which encouraged him to develop the idea further, leading to the thesis that all the proportions in the heavens could be explained by using aesthetically interesting combinations of geometrical figures to construct the circular motions found in the heavens.

Strange as it might seem that someone with astronomical gifts as great as Kepler's would expend energy on such a peculiar project, Kepler himself believed this kind of study to be an essential part of the program to improve astronomy: "Now by goodness, this a priori method improves the study of motion; and certainly we should hope for this improvement provided that others support us with the observations they have available" (Letter from Kepler to Maestlin, October 3, 1595; *KGW* XIII, 38). This way of conceiving of astronomy appealed to Kepler on a personal level as well, for it indicated that he could commune with God through the study of the heavens. This possibility was a great relief, given his initial desire to worship God through the study of theology: "I was determined to be a theologian: I was distressed by this for a long time. But look! Even in astronomy my work worships God" (Letter from Kepler to Maestlin, October 3, 1595; *KGW* XIII, 40).[3]

Maestlin, now Kepler's friend and mentor, helped Kepler revise, publish, and distribute the *Mysterium*. In the course of sending around copies of this book Kepler made contact with Galileo and, more important, Tycho Brahe (Galileo, after initially receiving the *Mysterium* with enthusiasm, was reluctant to correspond with Kepler). Brahe, on the other hand, though not in agreement with Kepler's thesis and method, was impressed and invited Kepler to work with him. The timing could not have been better. All Protestant teachers save Kepler had been expelled from Graz (it seems that Kepler's friend Herwart von Hohenburg had interceded on his behalf). The situation, nonetheless, was increasingly uncomfortable, as Kepler reported in a letter to Maestlin (August 19/29, 1599; *KGW* XIV, 56–58). Although he wished to return to Tübingen, Kepler feared that the political and religious climate would not allow it. But Emperor Rudolph II invited Brahe to take up the post of court mathematician in Prague, which he did. He settled in the nearby Castle of Benatky, which was close enough to Graz that Kepler felt he could make the journey. Brahe's invitation was Kepler's escape.

This situation, as unfortunate as it was for Kepler, gives us insight into his character that can aid the interpretation of his work. We see here, and elsewhere, that it ran counter to Kepler's character to lie or even hide his beliefs for the purpose of political expediency (especially important for the interpretation of his *Apologia*). Kepler could have avoided many of the political troubles he encountered were it not for the strength of both his convictions and his character. In a letter to Herwart von Hohenburg (a Catholic), he wrote, "I have not learned to publically put on appearances; I am serious in my religious practices, it is not just an act for me" (December 16, 1598; *KGW* XIII, 269).

Before Kepler could work with Brahe, he had to clear up a nasty piece of business. He had written a letter to Nicholas Reymers Baer (Ursus), the imperial mathematician, about his discoveries in the *Mysterium*. This letter was full of praise for Ursus (understandable given Kepler's youth and his awareness of Ursus's position). Ursus, without Kepler's knowledge, let alone permission, published the letter in 1597 in his *Astronomical Hypothesis*, where he advanced an attack on Brahe so vicious the censors would not approve it (he had to publish it himself; Rosen 1986, 14–15). As it turns out, Kepler's letter praised the world system that Brahe claimed had been stolen from him by Ursus. Kepler, now caught in the middle, and in need of a job and safe place to live, was forced to write the *Apologia* before he would be permitted to work with Brahe. He was loath to engage in a priority dispute and, as a means of accommodating both Brahe and his distaste, constructed the *Apologia* as a methodological treatise that only indirectly defended Brahe. As we shall see in chapter 3, Kepler managed to accomplish this political task while writing a treatise on methodology that he could and did stand by.[4]

This event, though unfortunate for Kepler, is extremely fortunate for the historian of philosophy of science because Kepler, though an astute thinker on methodological issues, preferred to address such matters only as they arose in the course of his work in natural philosophy. Thus, the *Apologia* is the only work of Kepler to address at length his views on the proper construction and testing of hypotheses. It is also an extremely important work for the history of astronomy, for here Kepler first argues at length for the necessity of grounding astronomy in physics, thereby laying the groundwork for the Keplerian revolution in astronomical method.

Unfortunately for the historian of philosophy of science, because Kepler was restrained to defend Brahe, one does not see in the *Apologia* an all-out methodological defense of the Copernican system. Kepler left out or left implicit any approach that would distinguish between the Brahean and Copernican system. Consequently, the *Apologia* should by no means be taken as a complete treatment of Kepler's position on the proper method of natural philosophy.

Kepler visited Brahe for a few months in 1600, a visit that ended acrimoniously (we do not know the exact nature of the argument). Brahe's suspicions that Kepler may have been in league with Ursus may explain the outcome—Rosen (1986, 258–63) cites two letters by Brahe written in March and April 1600—as well as the reason Brahe was so possessive of his observations. Shortly after, however, Kepler apologized and was invited to work with Brahe again. At the end of September 1600, Kepler moved with his family from Graz to Prague, and on October 24, 1601, Brahe died. Kepler was awarded Brahe's position and access to his observations—a rich treasure because Brahe, in addition to being a precise observer, also made extensive observations. On the basis of these observations Kepler undertook his study of the orbit of Mars. Originally, Kepler had intended to use Brahe's observations to buttress and further his argument in the *Mysterium*, but, as often happens, his work took a different turn: "I would have brought to completion my dissertation on the harmony of the world long ago already except the astronomy of Tycho took me over to such a degree I thought I would lose possession of my sanity" (Letter to Herwart von Hohenburg, July 12, 1600; *KGW* XIV, 130).

It is our good fortune that Kepler started working with Brahe's observations when Christen Sörensen Longberg, otherwise known as Longomontanus (one of Brahe's assistants), was working on Mars; indeed, Kepler entertained the possibility that the timing was due to God's intervention (*AN*, 185; *KGW* III, 109). The orbit of Mars deviated from the circle more than the other planets (except Mercury), which made it possible for Kepler to discover that the orbit was noncircular (even so, the deviation is not so obvious that it could be easily observed, as we shall see in chapter 4).

Kepler initially thought he would be able to solve the problem of Mars's orbit in eight days, but he ended up struggling with this problem for many years. He determined the orbit to be noncircular in 1602 and, in the same year, introduced the area rule—namely, that the radius vector sweeps out equal areas in equal times. By 1605 he discovered the first planetary law, that the orbit is an ellipse with the sun at one focus. The title he gave this work is *Astronomia nova ΑΙΤΙΟΛΟΓΗΤΟΣ, sev physica coelestis, tradita commentariis de motibus stellæ martis ex observationibus G. V. Tychonis Brahe*. As the title indicates, the foundation for his new astronomy is the methodological principle proposed in the *Apologia*, that astronomical theorizing is guided by physical reasoning.

To get this momentous work published took Kepler several years. He had to fight with Brahe's heirs over his right to use Brahe's observations (he indicated in a letter to Maestlin on Dececember 14, 1604, some of the difficulties he had with Brahe's son-in-law, Tengnagel; *KGW* XV, 73). Financial obstacles to the printing complicated matters further. But in 1609 one of the most important books in the history of astronomy was finally published.

Kepler also experienced financial difficulties in maintaining his domestic situation. Although he had been given Brahe's position, he frequently encountered problems in receiving his salary from Rudolph II (indeed, though his position began in October 1601, he did not receive his first payment until March 1602). Kepler reported in 1611 that the court owed him money and requested payment of 2,333 gulden (Letter from Kepler to Duke Johann Friedrich, March 9/19, 1611; *KGW* XVI, 368). On his appointment in 1601, he was to receive 500 gulden annually (Casper [1959] 1993, 123).

In 1612, after his wife Barbara's death, Kepler moved to Linz, where he stayed for fourteen years. During this period he wrote and published the *Epitome* and the *Harmonice*, the former a mature treatise on "Copernican" (actually, Kepler's own) astronomy, the latter an exploration of the divine and harmonious nature of the heavens. Kepler's Earthly experiences must have stood in marked contrast to his initial perception of the heavens. Shortly after arriving in Linz, he was excluded from Communion on the basis of his religious beliefs by a Lutheran pastor. Kepler was hurt and commented in a letter to Maestlin on December 22, 1616, that "he who accuses me of novelty-seeking on even the smallest matter causes me injury." Despite this, he stood firm: "For me, to put on an appearance in matters of conscience is not healthy" (*KGW* XVII, 203). Several years later his mother was tried for witchcraft, and he interrupted his work on the *Epitome* to intercede on her behalf. During his journey to visit his mother he read the *Dialogo della musica antica e moderna* of Vincenzo Galilei (Galileo's father) and, inspired by it, began work on the *Harmonice* (Letter to Matthäus Wacker Von Wackenfels, 1618; *KGW* XVII, 253). He had been working on the remaining books of the *Epitome* and the *Rudolphine Tables* (he had already published the first three books of the *Epitome*), but when his daughter Katharina died in February 1618, he turned to working on the *Harmonice*, which he found more soothing to his troubled soul: "I even entreat you, friends, do not damn me entirely to the earlier mathematical calculations [the *Rudolphine Tables* Kepler had been working on]. Grant to me the time for philosophical investigations, my only luxury" (Letter to Vincenz Bianchi, February 17, 1619; KGW XVII, 327).

Peace seemed determined to evade Kepler, and while attending to the *Harmonice*, the Thirty Years' War broke out (indeed, it did so in the same month, May 1618, that Kepler discovered the third planetary law). Despite these enormously unsettling events, Kepler finished and published the *Harmonice* and the remaining books of the *Epitome*.

Despite Kepler's search for peace in heavenly harmonies, he seemed to have found strife there as well. As I argue in chapter 6, Kepler was unable to construct a fully satisfying harmonic archetype, which must have troubled him enormously. One cannot help but wonder if Kepler found it fitting that even the heavens reflected the trouble experienced on Earth; he even remarked that, "The Earth sings MI FA MI, so that even from the syllable you may guess that in this home of ours MIsery and FAmine hold sway" (*HM* 440; *KGW* VI, 322).

In 1626 Kepler moved from Linz to Ulm and completed the *Rudolphine Tables*. More so than his earlier works, these tables secured his fame, for they were the most extensive and accurate astronomical tables yet produced. It was on the basis of these tables that many seventeenth-century astronomers read his other works (Russell 1964).

Kepler did not have many years left to live. He did, however, experience two personal events that brought him great happiness. One was the marriage of his daughter Susanna to Jakob Bartsch, a young student of astronomy and medicine who helped Kepler with the calculation of ephemerides (these are computed positions of the sun, moon, and the planets for every day of a given year). The other was the birth of another daughter by his second wife, Susanna Reuttinger (which he announced in a letter to Philipp Müller, April 1630; *KGW* XVIII, 429–30). In 1627 Kepler left Ulm and ended up in Sagan. Then, in 1630, Kepler started on a journey from Sagan to Linz to collect the money still owed him, and to give the ephemerides volume he had completed to Wallenstein (who had been Kepler's patron and friend from some years now). On a stopover in Regensburg, he took ill and died.

During his fifty-nine years, Kepler wrote many works that changed the face of astronomy and contributed greatly to optics, celestial physics, and what we now call philosophy of science, all while enduring personal and political hardships the likes of which it is difficult to imagine. Although I have barely covered the rich and remarkable texture of Kepler's life,[5] it should be clear why Kepler scholars tend to express not only deep admiration for the work but real affection for its author. As Caspar writes;

> It is the halo of his personality which draws many under his spell, the nobility of his character which makes friends for him, the vicissitudes of his life which arouse sympathy, and the secret of his union with nature which attracts all those who seek something in the universe beyond, and different from, that which rigorous science offers. In their hearts they all quietly bear veneration and love for this exceptional man. For no one who has once entered the magic sphere that surrounds him can ever escape from it. (Caspar [1959] 1993, 13)

Kepler's Astronomical Inheritance

To get a sense of the astronomy Kepler inherited, one needs to begin with Aristotle's celestial physics and Ptolemy's mathematical astronomy, for despite the developments in physics and astronomy over the centuries, the conception of the heavens passed on to Kepler was essentially that of antiquity.

Aristotle viewed the universe as divided into two main regions, the celestial and the terrestrial. He conceived the terrestrial to be a place of imperfection and change, composed of four elements (earth, water, fire, and air). Each of these elements moved in accordance with its natural inclination: earth and water are essentially heavy and so move downward; air and fire are essentially light and so move upward. The directions up and down were defined relative to the center of the universe, the Earth.

Celestial bodies did not fit into this terrestrial classification system because by all appearances they moved neither up nor down but in a circle. The heavens were believed to be the realm of perfection, and the circle and sphere the most perfect geometrical figures since they represent perpetual unchanging motion. It was obvious, then, to infer that the natural motion of heavenly bodies was circular. Moreover, because the essential nature of a body determines its natural motion, and because celestial bodies do not travel in any of the directions natural to the four terrestrial elements, they must be composed of a fifth (more perfect) element, one not found on Earth (Aristotle, *De caelo*, bk. 1, ch. 2). The natural physical explanation is that spheres exist in the heavens and rotate, carrying the bodies with them. Notice that in this system, the focus is not on the orbit of the observed body but on the mechanism that carries it around. The path of the body is an epiphenomenon. Not until Kepler did the planet's orbit become a phenomenon of principal interest.

The business of accounting for celestial phenomena by appealing to eternally rotating spheres would be a simple matter were it not for the irregular motion of the planets (which Ptolemy referred to as the five wandering stars). The planets have a daily westward motion with the stars, and a gradual eastward motion through the stars; for example, Mars takes an average of 687 days to complete its eastward circuit through the stars. But they also undergo periods of retrograde motion. Retrograde motion is when the planets slow down, then move backward (to the west), and then resume their course eastward.

Aristotle had inherited a very clever way of accounting for this motion using only concentric spheres (he cited Eudoxus and Callippus; *Metaphysics*, bk. 12, ch. 8, 1073b18–31). The basic model from Eudoxus has the planet placed on a sphere which turns on its axis (figure 1.2a). The poles of this sphere are embedded in another sphere, that rotates around a different axis (figure 1.2b), and the poles of the second sphere are embedded in a third sphere (figure 1.2c). The two innermost spheres, which rotate in opposite directions, produce a planetary path that resembles the infinity symbol, or a reclining figure eight. The third sphere carries the planet eastward through the stars, and a fourth sphere (not shown) carries the planet westward to produce its diurnal motion. Under the physical (metaphysical) constraint of uniform circular motion, one could still vary pole placement and circular velocity of the individual spheres to produce a variety of patterns.[6] In modifying this system, Aristotle added to Callippus's thirty-three spheres another twenty-two for a fifty-five-sphere model. Thus, though the physical principles were simple, the resulting model was quite complex.

This system had a powerful influence on physical and qualitative accounts of the heavens up to Kepler's time (although the greatest influence on mathematical astronomy was the Ptolemaic system). The physical principles had not gone unchallenged before Kepler, but they remained a part of the orthodoxy until Kepler finally rejected them. The first of these

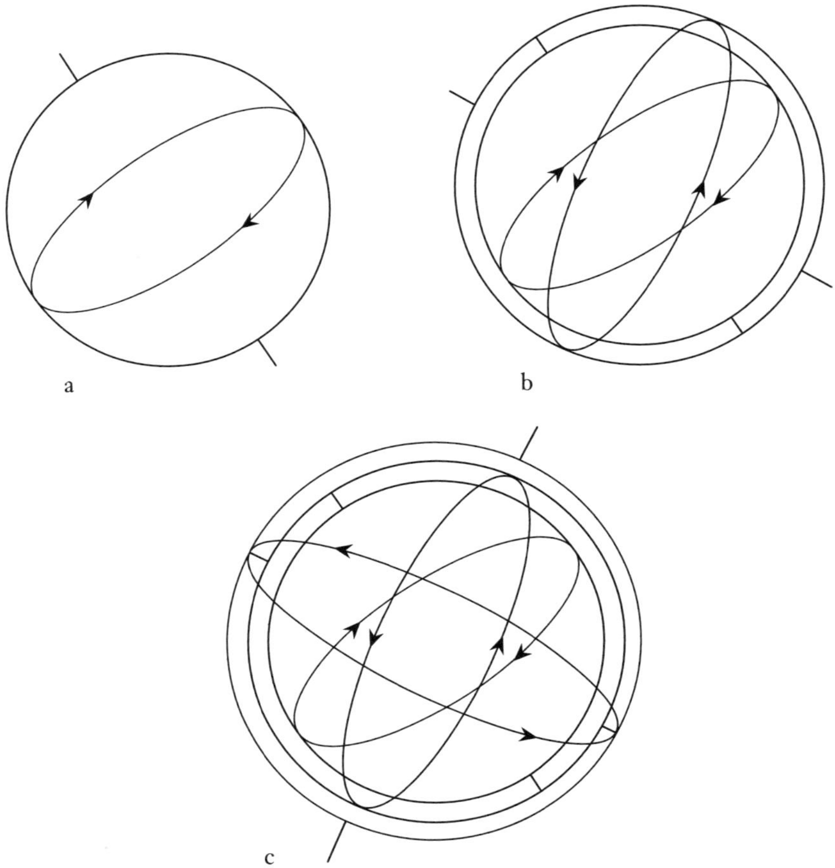

Figure 1.2. Eudoxus and the Homocentric Spheres

principles is the absolute distinction between the celestial and the terrestrial. Not only did the celestial move in a manner foreign to terrestrial bodies, but it was made of an altogether different kind of material. This principle left the door wide open for skepticism about our ability to know anything about the heavens. One could not use information obtained by studying Earthly objects in one's exploration of the heavens. According to Aristotle, celestial physics and terrestrial physics were two completely distinct disciplines (*Posterior Analytics* 79a8–12, 81b4; *Physics* 193b22–194b15; *Metaphysics* 1036a4–13). To complicate matters further, the author of the pseudo-Aristotelian *Mechanics*[7] (847a10–30) held that, although terrestrial physics was related to mechanics, they were not the same discipline; indeed, terrestrial physics was the study of natural motion and mechanics the study of unnatural motion.[8]

Second, the idea of perfect circular motion was such a compelling and beautiful idea that it seemed impossible to consider any other kind of geometrical figure in constructing planetary models. Later the idea was further entrenched by the Judeo-Christian conception of the universe as created by a benevolent God (who surely would choose to create the heavens to represent perfection). Moreover, by Kepler's time there was a long and *successful* tradition of representing planetary paths using only circles. Prior to Kepler's work in the *Astronomia*, there was no reason even to suspect that planetary motions were noncircular.

The third and related notion is that of the uniform velocity of the spheres. To conceive the sphere as rotating nonuniformly would be an anathema to the reasons for choosing the sphere in the first place, since the sphere represents eternal unchanging motion. Besides, to all appearances, the fixed stars rotate around the Earth uniformly. It was both reasonable and philosophically defensible to assume that all motions in the heavens are uniform.

Finally, it is important to note that, though this style of reasoning may appear to the modern mind as metaphysical, it was considered physical. The hallmark of Aristotelian physics was to infer what an object would do from the idea of its essence. Indeed, when Kepler argued that it was permissible to use physical considerations in astronomy, he appealed to Aristotle as predecessor (discussed in chapter 5).

Claudius Ptolemaeus (second century A.D.) was the first to leave us a detailed mathematical astronomy in a work originally titled *Mathematical Treatise* or *Mathematical Collection* but now more commonly known as the *Almagest*. The *Almagest* is a mathematically and empirically rigorous treatment of spherical astronomy, planetary theory, solar and lunar theory, and eclipses. For our purposes, a simplified discussion of Ptolemy's planetary theory will suffice.

In the *Almagest* Ptolemy developed the theory of epicycles—which he credited to Apolonius and other mathematicians in book XII, section 1 of the *Almagest*—to account for the retrograde motion of the planets. The basic epicyclic model employs a main circle, known as the deferent, and a secondary circle, known as the epicycle (figure 1.3). If the planet is placed on the epicycle at P, which rotates counterclockwise, and the center of the epicycle at point C also rotates counterclockwise, the path traced out by the planet will form loops. One can vary the relative sizes and velocities of the two circles to trace out a variety of figures.

While the epicyclic theory could account for the retrograde motion of the planets, there were still other irregularities that needed to be dealt with. The sun and the moon, though they do not retrogress, spend a

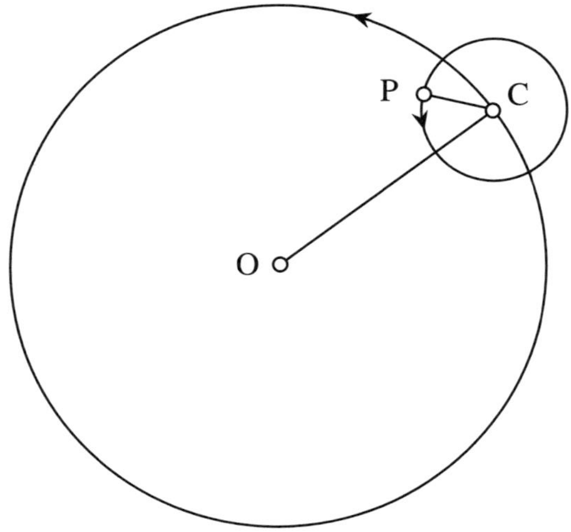

Figure 1.3. A Basic Epicyclic Model

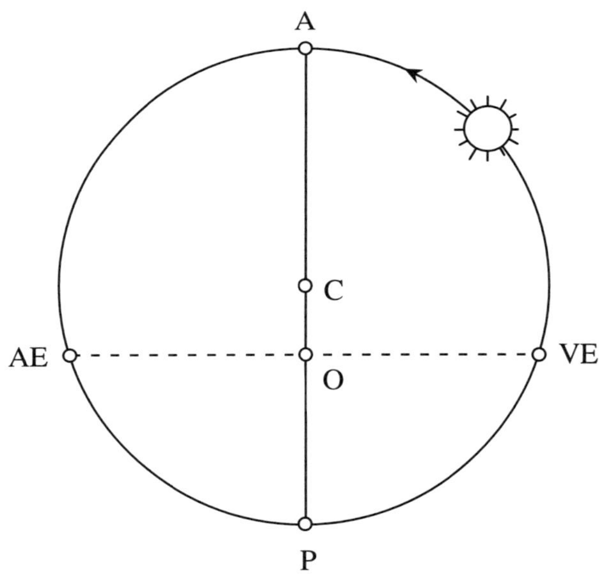

Figure 1.4. An Eccentric Circle

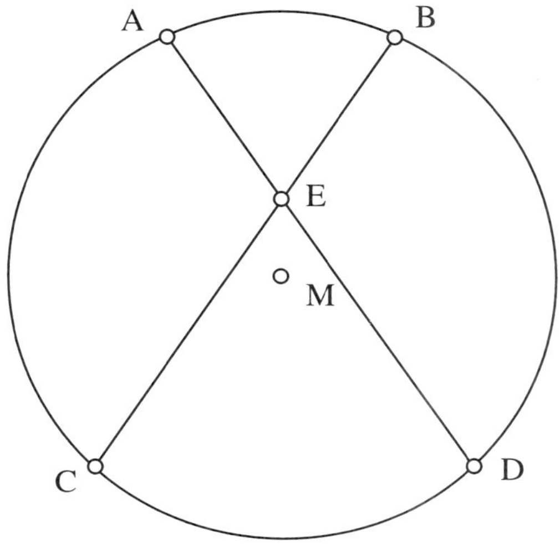

Figure 1.5. The Equant

greater amount of time on one side of the circle than the other. To save the principle of uniform circular motion, the main circle could be placed with the Earth off-center at O rather than at C (figure 1.4). This device, known as an eccentric, would explain why the sun spends extra time between the vernal (VE) and autumnal (AE) equinoxes. Another device is that known as the equant, where the angular velocity is uniform with respect not to the center but to some other point. In figure 1.5 the center of the circle is M, and the equant point, E. Angle AEB is equal to angle CED, but arc CD is larger than arc AB. So while the angular velocity is constant, linear velocity is variable. This has the effect of separating the center of uniform motion from the center of constant distances. The standard model in its simplest form used by Ptolemy to account for the planets was one with a "bisected eccentricity," which is where the Earth O, the center of the sphere M, and the equant point E are all placed on the apsidal line (the line between perigee [P] and apogee [A], which are the points on the circle least and most distant from the Earth respectively) such that the distance from the Earth to the center is equal to the distance from the center to the equant (figure 1.6).[9]

Strictly speaking, the Ptolemaic system was not fully compatible with the physical account of the heavens of the Aristotelian tradition, because it posited motions around centers other than the Earth as well as nonuniform motion. This tension gave rise to a number of different responses, helpfully taxonomized by Jardine (1984, ch. 7).

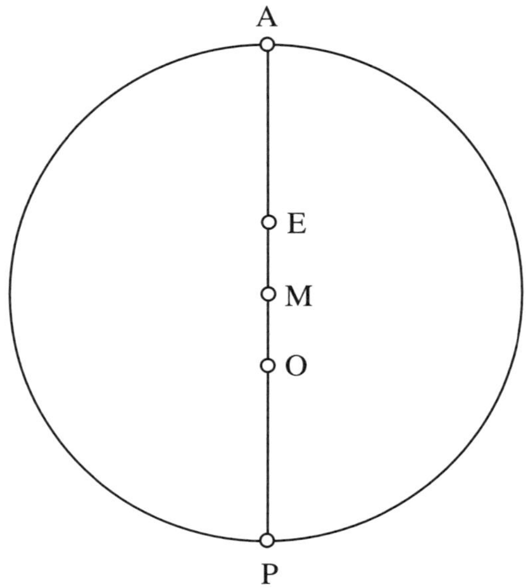

Figure 1.6. The Bisected Eccentricity

Some responded by providing rival cosmologies. Nicholas de Cusa (1401–1464), whom Kepler referred to as "divine," argued in his *De docta ignorantia* on Neoplatonic grounds that the Earth moves (bk. II, ch. XI). Key in his arguments is the role that aesthetic considerations of geometrical figures play (similar considerations occupy a central position in Kepler's cosmology). Girolano Fracastoro (1483–1553) and Giovanni Battista Amici (1511?–1538) challenged the Ptolemaic system by offering concentric theories—that is, geocentric theories devoid of epicycles and eccentrics (Jardine 1984, 231). Giovanni Pontano (1426–1503), in his *De rebus coelestibus*, rejected the idea of material spheres and proposed that the heavens are made of fluid, which the planets moved through "like birds in the air or fish in the sea." (This position was mocked by Kepler, who observed that the planets lacked the necessary appendages to swim or fly; *AN*, 377; *KGW* III, 237.)[10]

In a more conventional vein, Georg Peurbach (1423–1461) provided an influential account of the material nature of the Ptolemaic mechanisms in *Theoricae novae planetarum*. In this account, the epicycles were like globes enclosed within and carried by the rotating spheres. In figure 1.7, the center of the epicycle is at C, the planet at P, the center of the deferent at M, the equant point at E, and the Earth at O. Following up on Peurbach's work, Christoph Clavius (1537–1612) promised (but did not provide) a mechanical explanation of the equant (Lattis 1994, 69; notice that

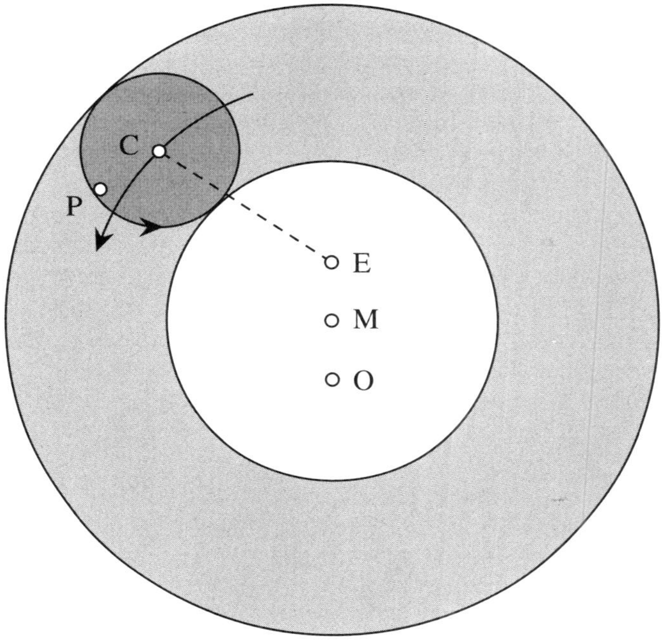

Figure 1.7. The Solid Spheres

this is a rival to the Aristotelian position since it runs contrary to concentrism). The solid-spheres models had been seriously challenged by the time Kepler arrived on the scene. Jean Pena argued that if there were solid spheres, light would refract in certain ways incompatible with observation (*Euclidis optica et catoptrica*, 1577; cited in Jardine 1984, 230). Tycho Brahe, based on his observations of the comet of 1577, argued to Kepler's satisfaction that one could determine empirically that the comet passed through areas where the spheres should have been (*Apologia*, 206–7; *KGW* XX1, 61–62).

Averroists like Alessandro Achillini (1463–1512), Agostino Nifo (*c.* 1469–1538), Girolamo Fracastoro (1483–1553), and Giovanni Amici (1511–38) claimed on the basis of Aristotle's *Metaphysics* and *De caelo* that Ptolemaic astronomy must be false, because it uses postulates other than concentric spheres (Jardine 1984, 231–32).

Others resolved the tension by denying astronomy *and physics* the ability to provide knowledge about celestial motions, on the grounds that the heavens are so remote and so foreign that we cannot hope to understand celestial causes. Jardine tentatively includes in this camp Giovanni Gioviano Pontano (1426–1503), Pierre de la Ramée (1515–72), Francesco

Patrizi (1529–97), Nicodemus Frischlin (1547–90), and Christian Wursteisen (1544–88) (Jardine 1984, 232–37).

Most sixteenth-century thinkers, however, took a more moderate, epistemologically cautious approach. Many were skeptical about the reality of epicycles and eccentrics and, on the basis of Aristotle's distinction between the subject matter of physics and that of mathematical astronomy, held that mathematical astronomy could not yield knowledge about the causes of motion (Osiander, the author of the anonymous "Letter to the Reader" introducing Copernicus's *De revolutionibus*, is one of the more famous authors to hold this position: Jardine 1984, 237–43). This position was not radically skeptical like the one mentioned in the preceding paragraph, because physics was still considered capable of providing some knowledge of the heavens.

This was the astronomy Nicholas Copernicus (1473–1543) inherited. My treatment of Copernicus will be very brief because my interest is more with the state of astronomy resulting from his contributions than with his grounds for rejecting geocentrism, how many epicycles were used in his system, whether his approach was radical or conservative, and so on.[11] As is well known, Copernicus's main contribution to astronomy was to liberate the Earth and set it into motion around a stationary sun. Under this system, the planets do not actually retrogress; rather, apparent retrogression is explained by the Earth passing (in the case of the outer planets) or being passed by (in the case of the inner planets) the other planets.

Although Copernicus's innovation was a significant achievement, the Copernican theory is at base a variation on Ptolemaic astronomy, albeit a radical one. An understanding of this point explains how Ptolemaic astronomy, though false, enjoyed such a high degree of predictive success. In figure 1.8a, the Earth is at O, and the planet at P (this is a figure for the outer planets). Under the Ptolemaic system O is stationary, the sun is somewhere on the line OS′, the distance between the Earth and the center of the epicycle at C is R′, and the radius of the epicycle is r′. Under the Copernican system, O now moves around stationary point S (the sun) with radius r, and the distance from the sun to the planet at P is R. Because the four lines form a parallelogram in which $r/R = r'/R'$ and the direction OP from the Earth to the planet is the diagonal, the Copernican system effectively exchanges the epicycle of the planet for the annual motion of the Earth. The two theories, in their simplest forms, are geometrically equivalent. In the case of the inner planets, the epicycle is exchanged for the orbit of the planet around the sun, and the radius of the deferent is exchanged for the distance from the Earth to the sun. In figure 1.8b, under the Ptolemaic system, the Earth is at O, the planet at P, the sun somewhere on the line OS′, the epicycle radius is r′, and the distance from the center of the epicycle to the Earth is R′. Under the Copernican system, the sun is at S, the Earth-sun distance is R′, and the sun-planet distance is r′.

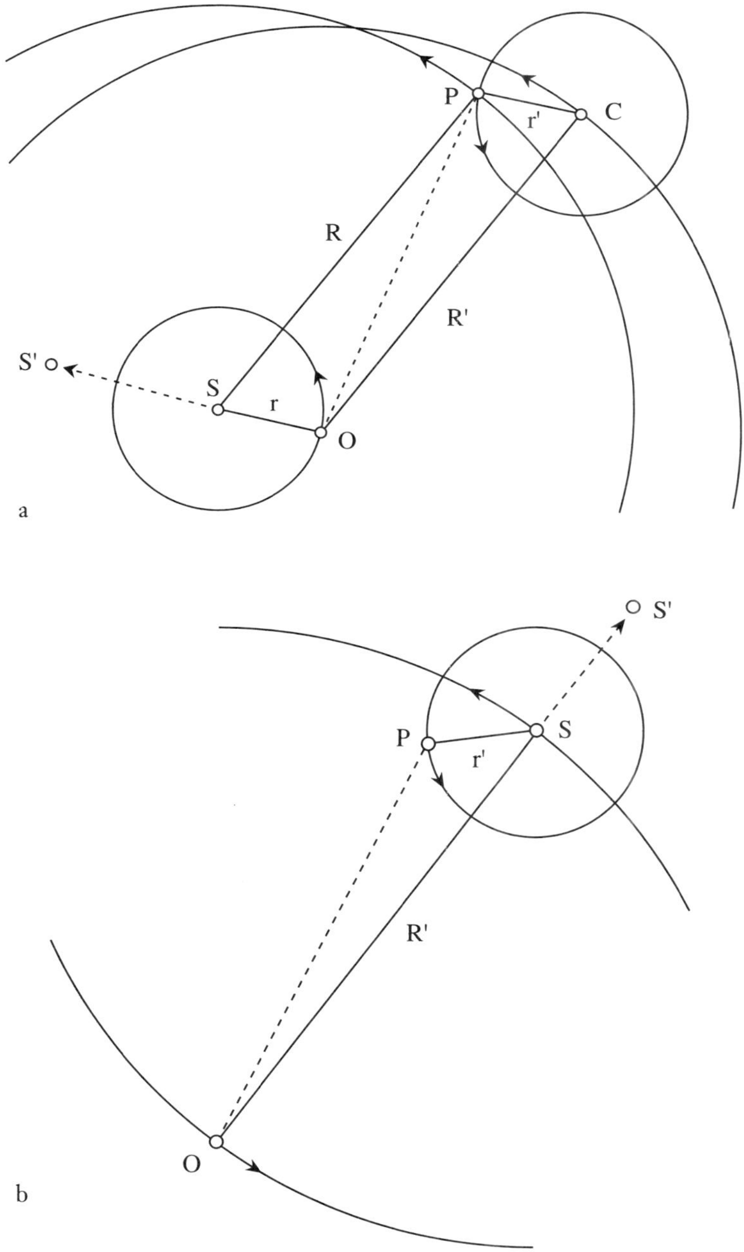

Figure 1.8. The Equivalence of the Copernican and Ptolemaic Hypotheses

Knowing the outcome of this story might mislead us into thinking that heliocentrism was the obvious rational choice, but Copernicus's theory faced obstacles.[12] First, we *see* the stars, the sun, and the planets moving around the Earth. Geocentrism has the advantage here in that it respects the appearances, whereas heliocentrism has to explain them away—a minor problem, of course, but there were other more serious problems for Copernicanism.

Given the empirical data then available, there were no compelling empirical grounds to prefer the Copernican system to the Ptolemaic. Indeed, there were empirical grounds for rejecting Copernicanism. Because under heliocentrism the Earth is moving around the sun, it is also moving relative to the fixed stars. Because the distance to the sphere of the stars was thought to be about 20,000 Earth radii, the stars ought to appear as though shifting from side to side. In figure 1.9, because the Earth travels from E_1 to E_2, the star S should look like it is moving from S_1 to S_2. But this effect was not observed. One could suppose, as Copernicus did, that the region of the fixed stars is sufficiently far away that such an effect would not be visible to the naked eye, but this raises another problem. Given the apparent diameters of the stars (believed to be 1' or even 2' for brighter stars), if they were as far away as Copernicus required (*De revolutionibus*, bk. 1, ch. 10), then they would be unbelievably large (over 200 times the size of the sun). Galileo was able to remove this objection when he noticed that stars observed through the telescope appeared much smaller than they did to the naked eye. But the appearance of the annual displacement of the stars was not correctly observed until 1838.

In addition, setting the Earth in motion raised all sorts of physical problems. How is it that heavy objects fall toward the center of the Earth if the Earth both rotates on its axis and around the sun? Why is it that we do not observe the effects of these two motions of the Earth on flying bodies, or on the behavior of clouds? Copernicus offered an Aristotelian answer. Because it is natural for the Earth as a spherical body to rotate, everything that belongs in the terrestrial region is carried along with the body of the Earth (*De revolutionibus*, bk. 1, ch. 8). This answer to the problem was reasonable, given that it was motivated by the conventional physics of that period in its appeal to notions of natural motion and the movement of bodies to their natural place.

One of the advantages of the Copernican theory is that one can use it to determine the order and distances of the planets without making arbitrary assumptions (*De revolutionibus*, bk. 1, ch. 10). Under the Ptolemaic system, for any individual planetary model, the *ratio* of the radius of the deferent to the radius of the epicycle can be determined empirically, but only individually, and the values of the radii cannot. More important, because there is no common unit of measurement between the different planetary mod-

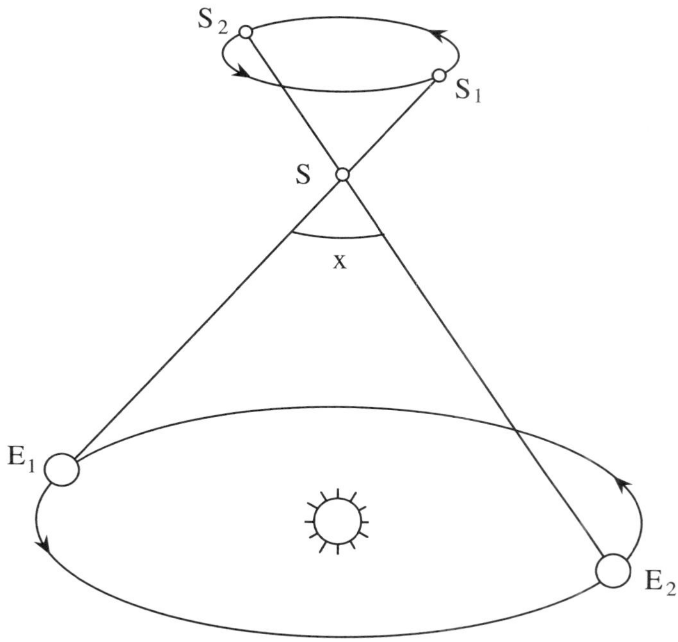

Figure 1.9. Stellar Parallax

els, one cannot determine empirically the relative distances of the planets. Under the Copernican system, the distances of all the planets are measured in terms of the radius of the Earth's orbit. In figure 1.8a, since the ratio of r:R can be determined, and since under the Copernican theory r (the radius of the Earth's orbit) is the same for all the planetary models, one can empirically determine planetary distances in terms of unit r.

This advantage could be turned against Copernicanism, however. Because the planetary distances under the Ptolemaic theory were not fixed, it was possible to model them such that the spheres of each planet would nest one on top of the other. This arrangement was physically plausible and metaphysically appealing. Under the Copernican system, the distances between the planets did not seem to follow any rule (the distances between the various planets were not equal, nor did they appear to increase or decrease in any mathematically sensible manner). The spheres would either have to have arbitrarily sized spaces between them, or would have to vary arbitrarily in thickness. Either alternative was implausible and aesthetically unappealing. Moreover, given that the world was assumed to be a created world, it would be difficult to explain why God would chose such an unattractive and arbitrary construction.

If that wasn't bad enough for Copernicanism, Tycho Brahe introduced a theory (in *De mundi aestherei recentioribus phaenomenis liber*) in which the Earth was stationary, the sun went around the Earth, and the planets revolved around the sun. For those wishing to maintain the Earth's stationary status, this alternative would have been appealing.

To make matters more complicated, the availability of more than one empirically adequate predictive system jeopardized the support that successful prediction might otherwise provide for any particular theory. The problem is this: if two contradictory hypotheses, at least one of which must be false, can make the same correct prediction, then predictive success cannot be used as an indication of the theory's truth (Kepler was to address this matter in the *Mysterium*, the *Apologia*, and the *Astronomia*).

Kepler came to realize that, given the lack of decisive empirical evidence for Copernicanism (and, indeed, some against it), the aesthetically unappealing Copernican distances, concerns about the plausibility of a moving Earth, and skepticism concerning astronomical hypotheses in general, any defense of Copernicanism would have to be on methodological, physical, or aesthetic grounds.

There were, however, considerable epistemological and conventional barriers to defending Copernicanism on extra-astronomical grounds. Westman (1980, 105–8, 133), in his excellent study of the institutional and social status of astronomers in the sixteenth century, examines the barriers to interdisciplinary work between astronomy, physics, and philosophy. He argues, citing evidence such as Osiander's anonymous "Letter to the Reader," that if physical conclusions are drawn from astronomical arguments, the structure of the liberal arts, as established by Aristotle, will be thrown into disarray. Aristotle claimed that each discipline has its own subject matter, which must be approached with the principles appropriate to that subject matter. In particular, knowledge was to be obtained through deductions based on essential properties delimited by and particular to each discipline. As a result, it was not obvious that one could draw physical conclusions from astronomical arguments, or astronomical conclusions from physical arguments.

There was some evidence for the legitimacy of Aristotle's position, at least as it pertained to astronomy. As already mentioned, since the advent of empirically adequate rival astronomical theories, it became clear that predictive success in astronomy could not by itself speak to the truth. The Aristotelian explanation for this problem is that because astronomy studies motion mathematically, and because the mathematical features of a dynamic system are accidental rather than essential properties of that system, and because knowledge is knowledge of essences, astronomy could not yield knowledge. This explanation was seized upon by many in the fifteenth and sixteenth centuries for it resolved the tension between astronomy and physics by reducing the former to a merely predictive discipline (more on this in chapter 5; see also Jardine 1984, and Duhen [1908] 1969).

In addition, the institutional status of the astronomer (and mathematicians generally) was relatively low in the sixteenth century. The universities did not have departments or chairs of astronomy or mathematics. Those who studied astronomy or mathematics did so in the process of seeking a degree in law, theology, or medicine (both Kepler and Maestlin studied astronomy as a part of their theological training; Westman 1980, 117–18). The social position of the mathematician was accordingly low and encouraged little beyond the provision of service instruction (Westman 1980, 120).

During this period, university reformers both Protestant (especially Melanchthon) and Jesuit (such as Christoph Clavius) were struggling to improve both the state of the art and the status of mathematical disciplines, including astronomy. At the same time there were increasing opportunities for astronomers to gain employment outside the university system (as court astronomer, for instance). Tycho Brahe made it clear that, unlike his position as court astronomer, a university position would place restrictions on his work that he could not tolerate (Westman 1980, 121–23). The astronomical milieu, then, was in flux, allowing greater opportunity for conducting innovative research.

Kepler's astronomical inheritance came most directly from Michael Maestlin (one sees this influence most clearly in Kepler's first chapter of the *Mysterium*). Maestlin was a well-respected astronomer and did significant work in promoting Copernicanism. Maestlin found that the trajectory of the comet of 1577 could not be satisfactorily resolved under a geocentric theory but could be under a heliocentric one. This result, coupled with the superior elegance and simplicity of the Copernican theory, persuaded Maestlin that it was true (Westman 1975, 333), and he impressed these arguments upon Kepler (*MC*, 63; *KGW* I, 7: elegance and simplicity are discussed in chapters 2 and 3). Maestlin also provided a most lucid account of planetary theory, which he appended to Kepler's *Mysterium*. This account considerably improved the Copernican theory, rejecting the planetary values given by Copernicus and using instead those Maestlin himself calculated from Erasmus Reinhold's 1551 *Prutenic Tables* (Grafton 1973, 523–32).

Despite Maestlin's interest in the empirical and mathematical aspects of Copernicanism, he had no interest in using physical arguments in this context. Indeed, on the grounds that such arguments in astronomy were methodologically improper, he chided Kepler for using them (Maestlin's letter to Kepler, September 21, 1616; *KGW* XVII, 186–88). So while Maestlin tutored Kepler in technical astronomy, and inculcated an appreciation of the mathematically harmonious aspects of the Copernican system, he neither inspired nor encouraged Kepler's unusual interest in examining astronomical phenomena physically. He did, however, support Kepler's interest in archetypally explaining the structure of the heavens, which, as we shall see, played a critical role in justifying Kepler's grounding of astronomy in physics.

Kepler's Philosophical Inheritance

It would be impossible to survey the wealth of philosophical activity in the Renaissance in such a short space, for these times were marked not only by political, religious, technological, and cultural change, but philosophical change as well. Numerous books have been written on this topic (Gilbert 1960; Schmitt 1975; 1983; and Kristeller 1979, to name a few), and rather than repeat their work, I focus briefly on some of the philosophical trends that are of particular relevance to Kepler's work. A fuller picture of these trends, and the bearing they had on Kepler's thought, will emerge in the course of this book.

One general philosophical problem of concern to Kepler was the proper method of natural philosophy. Kepler was not alone, however; during his time there was a lively interest in methodology and epistemology.

The classical account of method is in Aristotle's *Posterior Analytics*. As already mentioned, Aristotle recommended that a proper scientific account of natural phenomena was to reason philosophically from essences to effects.[13] Because the cornerstone of knowledge in natural philosophy is the essences of objects, knowledge about one class of objects is not obviously transferable to another class. Therefore, even if it seems obvious to the modern mind that natural and constructed objects are subject to the same laws of motion, under the Aristotelian conception, because these two classes of objects have different essences, they are not obviously subject to the same laws. By the same token, because celestial and terrestrial bodies belong to different classes of objects, the laws of terrestrial physics may not transfer to the study of celestial bodies. In addition, because the mathematical features of objects were not thought to be part of the essential nature of those objects, it was not obvious that mathematical methods could be legitimately used in physics.

Although the Aristotelian account of method had fallen under critical scrutiny by Kepler's time, and indeed had fallen completely out of favor with some—Francis Bacon, for example, accused Aristotle of contaminating natural philosophy with his logic (*Novum organon*, bk. 1, sect. 54)—it was still extremely influential, and (as I argue in chapter 5) Kepler took it quite seriously.

At Tübingen University, Aristotle's *Posterior Analytics* was taught by Andreas Planer (one of Kepler's teachers). Planer followed Aristotle in holding a priori demonstrations from causes to effects (based on the object's essence) in higher regard than a posteriori reasoning from effects to causes (Methuen 1998, 182). This conviction was echoed in Kepler's discussion of his achievement in the *Mysterium*. Kepler did disagree with Planer's Aristotelianism on one significant point; unlike Planer, Kepler believed that mathematics could legitimately and fruitfully be applied to other disciplines (Methuen 1998, 199), although, as we shall see in chapter 5, he could justify this application in an Aristotelian fashion. Both Maestlin and Georg Liebler, who taught Kepler Aristotle's *Physics* and *De caelo*, held that mathematics, the study of quantity, does not concern itself with motion or change, as does physics, and consequently the two disciplines are distinct (Methuen 1998, 194). Maestlin also objected to Kepler's merging of mathematical astronomy and physics from the other side. As mentioned earlier, he complained that, by using physics in astronomy, Kepler was not confining himself to the proper method of astronomy. Liebler, however, viewed mathematics and physics as addressing the same objects from different perspectives (Methuen 1998, 194), and this connection is very important for Kepler's work.

Equally important was the rising prominence of the mathematical disciplines. As already noted, the status of mathematics had been poor, and many felt that the mathematical arts were held in inappropriately low esteem (indeed, Kepler cited the poor regard for teaching astronomy as a reason for his reluctance to go to Graz). As a result, there was a kind of self-conscious mission to revitalize mathematics. In the sixteenth century, editors of Aristotle's *Posterior Analytics* began advising readers of the need to understand geometry in order to understand the *Posterior Analytics* (Gilbert 1960, 86–88). Practitioners of the mathematical arts were being praised by influential figures. Melanchthon, for example, hailed Peurbach and Regiomontanus as "restauratores" of the mathematical arts (cited in Jardine 1984, 264). University reformers worked hard to raise the status of mathematics and to improve the position of mathematicians in the university system. This mission continued into the seventeenth century, when many, like Brahe in *De disciplinis mathematicis oratio*, introduced their treatises with discussions on the excellence and certainty of mathematical method, and its practical utility in other disciplines (in Brahe's case, this work primarily concerns astrology).[14]

Kepler encountered Tübingen in the middle of this transition. Although mathematics was still seen as subordinate to theology, many of the faculty were beginning to hold it in higher esteem. Like Brahe, Planer approved of the certainty of mathematics proofs, though he did not believe that such proofs should be used in other disciplines. Liebler, however, lauded Jacob Schegk (who was at Tübingen from 1536 until 1577) for applying mathematical proofs to other areas of philosophy (Methuen 1998, 199). Simon Grynaeus, who was in Tübingen from 1534 to 1535, believed that the clarity of mathematical proofs could aid in the interpretation of observations (Methuen 1998, 168–69).

The mechanical arts were also attracting attention, partly because mechanics was a successful discipline, and partly because it was a physical discipline that made effective use of mathematical demonstrations.[15] Kepler was greatly impressed by the mechanical arts, and he proposed that the world is a machine to be explained by mechanical principles.[16]

There was also a renewal of interest in Platonism and Neoplatonism, a result in part of the Renaissance commonplace that Plato was the patron of mathematics. Also, Plato's conception of a transcendent unchanging reality (the doctrine of the forms) was taken up by a number of thinkers, including Nicholas de Cusa (1401–64) Marsilio Ficino (1433–99), Pico della Mirandola (1463–94), and Robert Fludd (1574–1637).[17] Kepler himself called Plato and Pythagoras his true masters (in a letter to Galileo, October 13, 1597; *KGW* XIII, 145; translated in Baumgardt 1951, 41), presumably on the basis of the shared belief in a mathematical metaphysics. The Neoplatonic doctrine of emanation (that the one true reality radiates from itself all aspects of reality and is thus reflected in all things) in the writings of Plotinus (A.D. 205–70), Proclus (A.D. 410–85), and later Nicholas de Cusa was also influential and is evident in Kepler's thought.

Kepler was particularly impressed by Proclus's Platonic position that the mind is created with the imprint of mathematical ideas. Kepler's interpretation of Proclus was distinctly Christian, as we can see from this passage that Kepler quoted, adding his own clarifications in brackets, from Proclus:

> *Then it will have to be supposed that the soul itself is the generator of mathematical species and concepts. But if, containing them in itself as first patterns or paradigms, it makes them take their essential character, in such a way that their generation* (the Christian understands, the creation of sensible things) *is nothing but the propagation of species which were previously in it* (that the mathematical reasons for the creation of bodies were coeternal with God, and that God is pre-eminently soul and mind, whereas human souls are images of God the Creator, even in essentials in their own way, is known to Christians) *then we shall agree with Plato, in saying this, and the true essence of mathematics will have been discovered by us.* (*HM*, 299; *KGW* VI, 219; emphasis in original).

Kepler also agreed with Proclus that mathematics, by revealing the structure of our thought, could, in virtue of the doctrine of emanation, reveal the structure of the world. This idea was so important to Kepler that he opened the *Harmonice* with a quotation from Proclus's commentary on Euclid's *Elements*:

> (Mathematics) contributes things of the greatest importance to the study of nature, both revealing the orderly nature of the reasoning, in accordance with which the WHOLE has been constructed, and so on, and showing that the simple and primary elements, by means of which the whole of the heaven was completed, having taken on the appropriate forms among its parts, are connected together with symmetry and regularity. (*HM*, 7; *KGW* VI, 13)

Kepler called Nicholas de Cusa divine for asserting that the figure of the sphere represents God (*MC*, 93; *KGW* I, 23). It is not clear whether Kepler's interpretation is the same as Cusa's; Cusa claimed that the center and the circumference are one and the same in God (*De docta ignorantia*, bk. II, ch. 12). Kepler tended to speak of the divine sphere in spatial terms, but because the three parts of the sphere represent the Trinity, it may well be that in God the center and circumference are the same thing.

In her study of the curriculum at Tübingen and the works of Kepler's professors, Methuen (1998, 221) observes that, given Kepler's Neoplatonic inclinations, there was only a modicum of interest in Platonism and Neoplatonism at Tübingen. Such interest was not completely absent, however. Liebler paid attention to Plato and the Pythagoreans, Planer discussed both Plato's *Timaeus* (which Kepler viewed as a commentary on Genesis in the Bible: *HM*, 301; *KGW* VI, 221) and the theory of forms, and Grynaeus wrote a preface to Proclus's Neoplatonic commentary on Euclid (Methuen 1998, 99, 204). Kepler also was familiar with Nicholas de Cusa by the time he wrote the *Mysterium*. While the Neoplatonic movement in Italy was growing, there did not seem to be much awareness at Tübingen of the Italian Neoplatonists, with the exception of Pico della Mirandola (whom Liebler mentioned; Methuen 1998, 222).

Kepler did not embrace the Neoplatonic movement uncritically. For example, he approved heartily of Cusa's conception of a world structured by geometrical aesthetics, but took vehement exception to Fludd's Neoplatonic metaphysics, in part on the grounds that it was so transcendent as to have nothing to do with the empirical world (*Pro suo opere harmonices mundi apologia*). It was of the utmost importance to Kepler that theories be empirically respectable.

These trends—the legacy of Aristotle, the increased use of mathematics in physics, and rising interest in mechanics and Neoplatonism—did not fit easily into a coherent, programmatic system. For the natural philosopher wishing to draw from all of these, as Kepler did, the tension inherent in this complex of influences cried out for a solution. As a result, Kepler's work in physical astronomy is a rich resource for the history of methodological thought.

At the same time there was a renewed concern with observations, not just with making them, but with what to make of them. There are several issues here, all of which challenge the inference from observation to knowledge.

First, there are the commonplace optical illusions that we learn to deal with as infants. A table does not change in shape while being viewed from different angles. People do not grow in size as they approach. These ordinary sorts of illusion do not overly concern us, but they can be used, as John Locke and George Berkeley did, to drive a wedge between appearance and reality. Kepler understood Patricius to have claimed in contrast, that such illusions are in fact veridical, and judgments to the contrary false. For this he chided Patricius:

> [O]ne who paid heed to him could not move a foot without granting that a miracle occurred. For where, walking through the fields, he encounters hedges and things near to his path, he would believe, on the testimony of the sense of sight, that distant mountains are really following. He would not use the judgement of reason, nor even consult the common sense. (*Apologia*, 155; *KGW* XX1, 29)

These ordinary illusions thus called to Kepler's attention the need to consider how it is that we interpret observations, and the conditions under which our interpretations are veridical.

The problem of the interpretation of observations becomes more complicated in the context of astronomy. The motions of the planets, for example, appear chaotic and disorderly (they slow down, speed up, go eastward, then westward), but it was taken for granted by most that this could not be how things really are. The world, surely, could not be as chaotic as it appeared. The resulting problem is both epistemological and metaphysical. First, how can one access how things really are? Does one transcend (ignore) appearances, as Fludd believed, or can one find an underlying order that allows the appearances to be predicted and explained? Second, how should one conceive of the relationship between appearance and reality? Is there such a relationship, and, if so, is the world of appearance an imperfect reflection of reality, or does reality produce appearances in an orderly and discoverable fashion? Kepler believed that the apparent disorder of the universe is produced in an orderly fashion by simple underlying principles. The job of the natural philosopher, then, was to uncover these principles and show how order explains the appearance of disorder. Hallyn draws an analogy between this belief of Kepler and mannerist aesthetics. Under the mannerist conception of art, what was aesthetically pleasing is to express how the monstrous and chaotic can be produced by a simple ordering principle (Hallyn [1987] 1990, 160).

Second, there was a growing concern with observational error. Nicholas de Cusa claimed that we are likely to make errors in measurements, in part because our expectations will lead us to see things in one way rather than another (*De docta ignorantia*, bk. II, ch. X). Francis Bacon also warned that the expectation of order will lead us to see things as more orderly than they really are (*Novum organon*, bk. 1, sec. 45). Both adduced astronomy as an example of where the problem of observational error is especially keen.

Third, there was growing concern with the question of whether natural philosophers should process data or merely collect them. Bacon is often cited as commenting on what he saw as the three approaches to data: one can simply collect data, as an ant collects food; spin theories out of one's mind, as a spider spins webs; or take the data and process them, adding something to them in the process, as a bee makes honey. Bacon advocated the third way, the way of the bee (*Cogitata et visa* [1607] VI, 616). I do not know whether Kepler read *Cogitata et visa*; yet he too argued that the mind should impose a rational order on data.[18]

Fourth, if the mind *processes* data, then this raises the question of whether our conception of reality can be true (if the mind simply reflects like a mirror, then the answer is obvious). One answer to this problem was to hold, as many did in Kepler's time, that the ultimate structure of reality is imprinted on the human mind (as a result of our creation in God's image, or more Stoically, our being microcosms in sympathy with the macrocosm).

Fifth, there was the question of whether it is legitimate at all to move beyond observations, that is, to postulate hypotheses about what causes observable phenomena. In this connection, Pierre de la Ramée (also known as Ramus, 1515–72) offered his professorship as a reward for anyone who could construct an astronomy without hypotheses (*Prooemium mathematicum*). In the *Astronomia*, Kepler cited as unfortunate the fact that Ramus was no longer alive, for he could have claimed Ramus's chair. By this Kepler meant that he had indeed constructed an astronomy without hypotheses, where a "hypothesis" is understood to be something completely unwarranted, rather than about the causes of phenomena.

Also, as already mentioned, in astronomy there was a wedge between the truth of a theory and its predictive success, a problem highlighted by the availability of more than one empirically adequate theory. If true observations can be derived from false theories, then it is not obvious in what sense observations support theories. This problem, perhaps more than any other, spurred Kepler to develop a methodological account that detailed the structural features a theory must have in order to be supportable by observation.

The various positions possible on these issues surrounding observations paved the way for a variety of methodological approaches and epistemological stances on the kinds of conclusions that can be drawn legitimately from the a priori and a posteriori. We will see that in attempting to solve these problems, Kepler had much in common with important figures in early modern rationalism (in the final chapter I discuss some of the explicit and implicit references to Kepler in the writings of Descartes and Leibniz).

Kepler's proposal, in broad strokes, was that one can access truth by moving dialectically between the a priori and the a posteriori. The manner in which he did this was something of a novelty. As Maestlin wrote:

> The topic and the ideas [of Kepler's *Mysterium cosmographicum*] are so new that up to now they never entered anybody's mind. . . . For who ever conceived the idea or made such daring attempt as to demonstrate *a priori* the number, the order, the magnitude and the movements of the celestial spheres . . . and to elicit all this from the secret, unfathomable decrees of Heaven! (Letter of Maestlin to Hafenreffer, June 1596; translated in Baumgardt 1951, 37)

The reaction to this approach was mixed. Johannes Praetorius (a professor from Altdorf) believed that such things belonged to physics rather than astronomy. Georg Limnäus (from Jena), by contrast, was pleased that someone was approaching astronomy in a Platonic manner (Caspar [1959] 1993, 69). Brahe was dubious that the a priori could improve an a posteriori discipline: "If astronomy will be restored through prior reasoning on the dimensions of regular solids, rather than from embracing accurate *a posteriori* observations . . . we will anticipate this for far too long, if not altogether perpetually in vain before this task will be discharged" (Letter from Brahe to Maestlin, April 21, 1598; *KGW* XIII, 204).[19] Yet later we find Descartes yearning for an a priori foundation for the study of the heavens, without which he thought any theorizing would be uncertain (Letter of Descartes to Mersenne, May 10, 1632; Kenny 1970, 23–24). Later still, Leibniz argued that physics must have metaphysical foundations. But the seeds for the rationalist use of the a priori in empirical studies had already been sown, and Kepler's rationalist commitments were instrumental in his construction of the new physical astronomy.

2

The *Mysterium cosmographicum* and Kepler's Early Approach to Natural Philosophy

In 1621, looking back over an impressive career, Kepler commented that, "almost every book on astronomy which I have published since that time could be referred to one or another of the important chapters set out in this little book [the *Mysterium cosmographicum*], and would contain either an illustration or a completion of it" (*MC*, 39; *KGW* VIII, 9).[1] Because Kepler viewed the *Mysterium*, his first book, as the genesis of his later astronomical works, I begin my discussion of the relationship between Kepler's philosophy and his new astronomy here. The *Mysterium* was written as an extended defense of Copernicus at a time when the empirical evidence did not favor the Copernican hypothesis over the Ptolemaic. Consequently, the defense had to be nonempirical, and Kepler based it on aesthetic, metaphysical, and methodological grounds.

To the modern reader the *Mysterium* might seem a strange book indeed, belonging only in a library of curiosities. But it is important to bear in mind that this book was well received at the time of its publication and popular enough to warrant a second edition. It even impressed Tycho Brahe, prompting him to invite Kepler to work for him.

The Content of the *Mysterium cosmographicum*

The *Mysterium* is primarily about Kepler's notion of archetypes. Kepler believed that God created the world according to a plan. This plan, or pattern, is the archetype for the structure of the world (much like an architectural blueprint). Kepler believed, following the Pythagoreans and Plato, that this pattern was mathematical in nature and aesthetic in character. Because God created the physical world as a representation of this pattern, an archetypal model explains why the universe is the way it is. To use Aristotelian language (as Kepler often did), archetypal explanations provide an account of formal and final causes, rather than efficient or material causes (where an object's formal cause is its essential shape, its final cause is its purpose, and its efficient cause is the mechanism of its creation).[2]

Kepler explained the stationary aspects of the heavens (the sun, the fixed stars, and space) archetypally by drawing an analogy with the Holy Trinity. The idea was that the physical universe, in the shape of a sphere, is a unity with three parts, as is the Trinity. Via the geometrical analogy, then, the universe represents the Trinity: "the Sun in the center, which was the image of the Father, the Sphere of the Fixed Stars, or the Mosaic waters, at the circumference, which was the image of the Son, and the heavenly air which fills all parts, or the space and firmament, which was the image of the Spirit" (*MC*, 95; *KGW* I, 24). We see this move between the divine, the geometrical, and the physical repeatedly in Kepler's work. Although it is not clear in this passage how the Trinity explains the stationary aspects, earlier Kepler was more explicit in a letter to Maestlin, saying that like God, the sun is an unmoved mover (October 3, 1595; *KGW* XIII, 35).[3] This claim is not too surprising, given Kepler's Neoplatonic leanings. Neoplatonists like Plotinus (whom Kepler frequently referred to, especially in his letters)[4] "regarded light as a central feature of the world—at once a transcendental reality and a physical agent, one of the fundamental principles of cosmogony and epistemology, the source of life and movement, and a powerful theological symbol" (Lindberg 1986, 9). Because the sun was seen as the main source of light in the universe, Kepler was understandably inclined to ascribe to it the primary motive force as well.

To explain the moving bodies, Kepler hypothesized that the five regular geometric solids (the cube, tetrahedron, dodecahedron, octahedron, and icosahedron), nested inside each other in an order determined by their importance, explain not only the sizes of the planetary orbits, but why there are exactly six planets. Spheres circumscribing the perimeters of each solid correspond to the relative sizes of the planetary orbits. Because there are only five regular polyhedra, only six spheres can be inscribed (there is a sphere surrounding the nest, which is why there are six spheres rather than five), thus explaining the number of planets. Figure 2.1 depicts, clockwise from the upper left hand, the cube, tetrahedron, octahedron, dodecahedron, and icosahedron; figure 2.2 (from *KGW* VIII, between pps. 48 and 49) depicts the nested-sphere model with which Kepler enthusiasts are no doubt familiar. The use of the polyhedra to explain the nature of physical objects was borrowed from the ancients. In Plato's *Timaeus*, which Kepler referred to as "a commentary on the book of Genesis" (*KGW* VI, 221), the polyhedra were used to account for the nature of the basic elements.[5]

Kepler ordered the polyhedra on the basis of certain relationships to the sphere, on the assumption that the distinction between the sphere and other geometrical figures reflects the distinction between the Creator and the created (*MC*, 95; *KGW* I, 24). Next to the sphere, then, the more a figure resembles the sphere, the more perfect it is. The sphere is the simplest figure in that all points on the face are equidistant from the center,

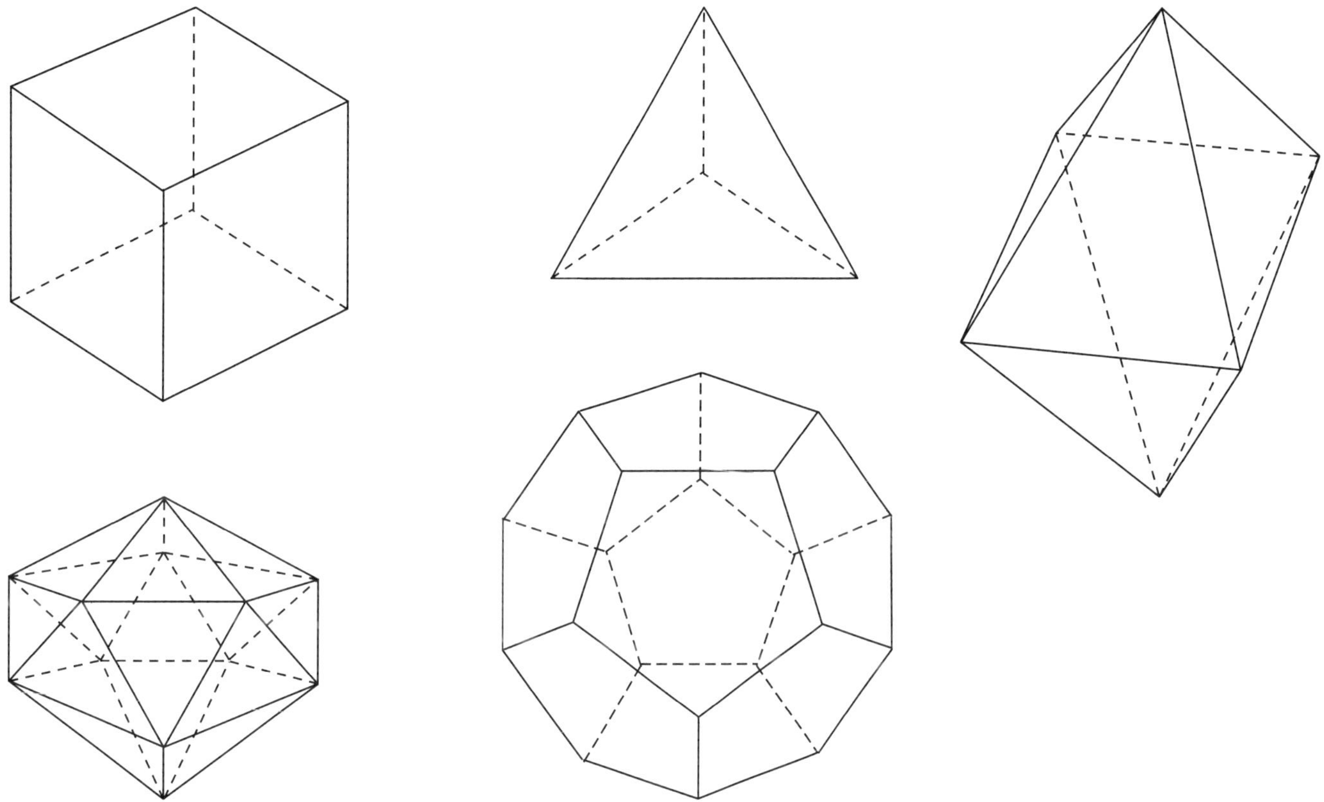

Figure 2.1. Regular Solids

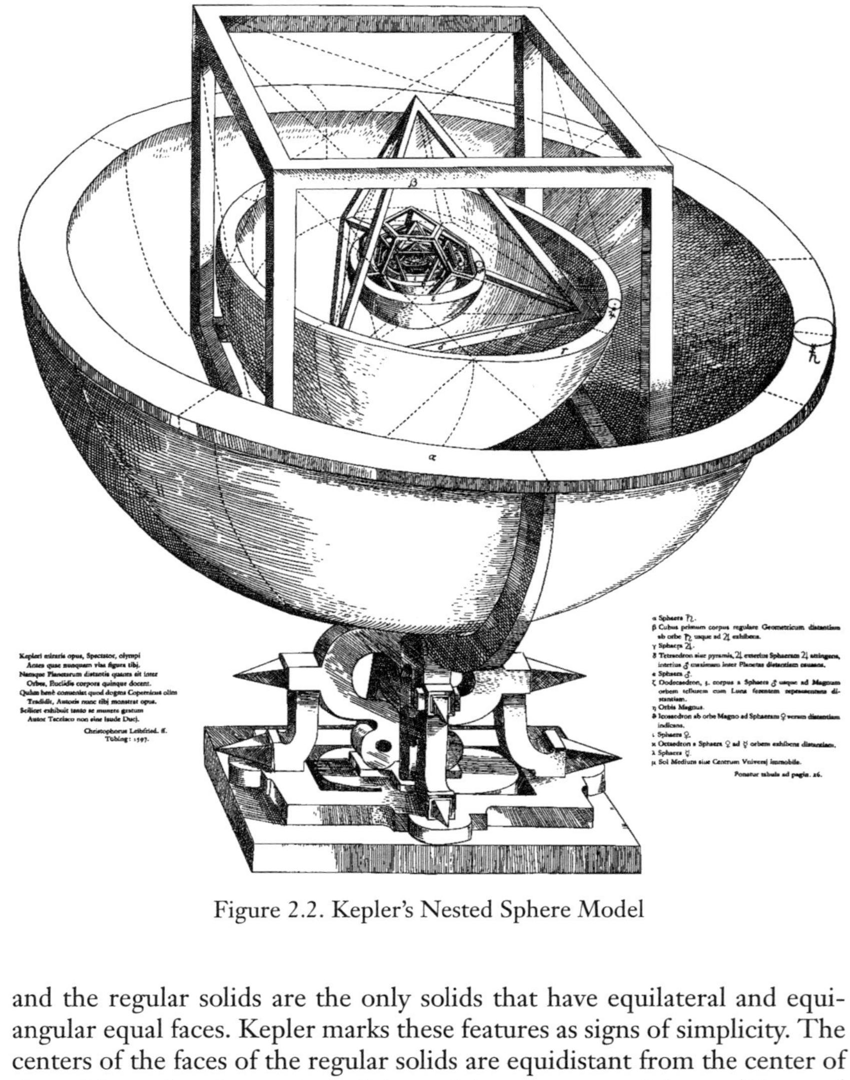

Figure 2.2. Kepler's Nested Sphere Model

and the regular solids are the only solids that have equilateral and equiangular equal faces. Kepler marks these features as signs of simplicity. The centers of the faces of the regular solids are equidistant from the center of the solids, and so the regular solids participate in some of the qualities, and hence some of the prestige, of the sphere. These features mark the polyhedra off from other solids, thereby limiting the number of solids to be ordered (*MC*, 97, 101; *KGW* I, 25, 27–28).

Kepler divided the solids into two classes. The primary class contains, in order of importance, the cube, the tetrahedron, and the dodecahedron. The secondary class contains, also in order of importance, the octahedron and the icosahedron. Kepler grouped and ordered the solids in chapters 3 through 7 of the *Mysterium*. Because thorough summary of Kepler's ordering principles is provided by Field (1988, 53–60), I restrict myself to a partial survey of Kepler's reasons for the ordering of the solids as he did. The primary and secondary solids differ from each other in various ways. For example, the secondaries share the same type of face, a triangle, whereas the primaries each have a different type of face. The vertices of the primaries join three faces, the minimum number for the formation of a solid, whereas those of the secondaries join four or five faces. Furthermore, the primary solids stand on a face, whereas the secondaries stand on a vertex: "For if you roll the latter [secondary] onto their base, or stand the former [primary] on a vertex, in either case the onlooker will avert his eyes at the awkwardness of the spectacle" (*MC*, 105; *KGW* I, 29).

This statement is rather puzzling, for it is not obviously offensive, or unaesthetic, to stand any solid on either its face or its vertex. Field (1988, 55) interprets Kepler to have meant that the appropriate standing of a solid depends on which standing gives it "rotational symmetry." The primary solids rotate harmoniously only if the axis is drawn through the center of a face, whereas the secondary solids require an axis through the vertices. Hence, when motionless, the primaries should stand on a face and the secondaries on a vertex. This passage highlights the importance for Kepler of not affronting aesthetic sensibilities. For Kepler, an offense against aesthetics is an offense against truth, for the world was ordered by God in an aesthetically pleasing manner.

Figural composition also plays a role in Kepler's ordering. The cube is the highest solid in part because it is the only solid with right angles, and because it is the only solid that can be generated from its face. The other figures are either compound or constructed from the cube (*MC*, 109; *KGW* I, 31). The tetrahedron follows the cube in part because it, along with the cube, generates the other figures. The remaining primary is the dodecahedron (*MC*, 111; *KGW* I, 33). The octahedron is the more important of the two secondaries because it is more similar to both the cube and the tetrahedron. The remaining secondary is the icosahedron (*MC*, 113; *KGW* I, 33).

The relative nobility of the regular solids determines their order in the nested model. The Earth's orbit holds a unique position, dividing the primary and secondary solids. The order of both primary and secondary solids is of ascending importance away from the Earth, the primaries toward the fixed stars, the secondaries toward the sun. The primaries are outside the Earth because "containing is proper for the more perfect" (*MC*, 107; KGW I, 30). The order from the sun outward, then, is the octahedron, icosahedron, dodecahedron, tetrahedron, and, finally, the cube. The result is six orbits (the orbit of Mercury is between the sun and the octahedron; *MC*, 105–14; *KGW* I, 30–34).

It strikes one that Kepler's ranking of the polyhedra was not based solely on a priori grounds. Plato also ordered the polyhedra on the basis of figural composition, but he considered the tetrahedron most basic since it could be used in the construction of the cube (*Timaeus*, 54–55d). Also, given Kepler's willingness to defy the authority of the ancients, he need not have based his ordering on the principle of figural composition. He could have ordered the solids on the basis of other spherical features. For example, the icosahedron has more faces than the other polyhedra, and hence is more like the sphere in the sense that it has more points equidistant from the center than any of the other solids.

Given the position of utmost importance the Trinity analogy affords the sun, it also seems plausible to suppose that the order of polyhedra could just as well be reversed, with the most important figures closest to the sun.[6] Despite Kepler's assertion that it is better to contain than to be contained, and despite the fact that the outer planets were generally considered superior (*MC*, 107; *KGW* I, 30), these assertions by themselves do not fully explain Kepler's particular ordering. After all, the most important body, the sun, is contained by all the polyhedra.

Kepler, of course, was not trying to determine the order of the planets, but rather to justify the Copernican hypothesis, which he had already accepted. The justification draws from both empirical and aesthetic resources. Kepler first put the spheres in order on the basis of rough empirical data (*MC*, 99; *KGW* I, 27),[7] and then constructed an aesthetic explanation of that order. Kepler believed that if one could construct an account on both empirical and aesthetic bases, it would likely be correct.

The empirical and the aesthetic play two different roles in Kepler's thought. The empirical determines a model's candidacy for explaining the world, and the aesthetic explains the final and formal causes of its structure. Because the universe is as it is for a reason, if a model is aesthetically preferable to its alternatives, one is warranted in holding that the model is not merely empirically adequate but explanatorily successful.

Given, however, that Kepler was using his polyhedral model to justify results he already accepted, he was concerned to show that his model was not ad hoc. This is why his extended defense of the model is based on a priori rather than a posteriori grounds: "Now, it might seem fortuitous, and not the result of any cause, that the six spheres of Copernicus accept these five solids into the spaces between them, if their actual pattern were not the pattern in which I have placed them" (*MC*, 105; *KGW* I, 29). We will see this pattern again in Kepler. Where Kepler speaks as though he derived the archetypal theory by a priori reasoning and then tested for fit against empirical data, it is more accurate to say that he first had the empirical pattern and then tried to fit to it an archetypal account. I take it that Kepler was insisting on the importance of the internal or theoretical plausibility of a hypothesis in addition to testing it for fit against either data or other theories. What Kepler seemed concerned with was the possibility of illegitimately manipulating hypotheses to fit the data—hence his insistence that the "sphere of Copernicus accept these five solids" in the order he determined a priori. If, however, there is a tight fit between an internally plausible hypothesis and the data, the proposed order is likely the result of a cause—that is, a probable account of reality.

After arguing for the polyhedral model on empirical and archetypal grounds, Kepler connected the polyhedra to astrological and musical concepts, a connection strikingly reminiscent of Ptolemy's between astronomy, astrology, and musical harmony. Kepler referred to Ptolemy's *Harmonics* in the *Mysterium*, but did not have access to it at the time. (He requested the book from Herwart on September 14, 1599; *KGW* XIV).[8] Kepler's comments on harmony in the *Mysterium* were not forerunners to his *Harmonice*. As Stephenson (1994, 89) notes, Kepler's early archetypes were exclusively geometrical. Kepler's purpose was to show that the pattern that he saw in planetary movements was connected to other patterns, hence increasing the likelihood that he had found an archetypal pattern, a divine blueprint for the universe (*MC*, chs. 9–12).

At this point Kepler considered the polyhedral hypothesis probable but in need of more empirical support (*MC*, 149; *KGW* I, 43–44). He then returned to a more careful analysis, using the polyhedral model to generate distances to be tested against distances calculated from Copernican data. The fit was close enough, giving further support to Kepler's confidence. The following is Kepler's table in chapter 14 (Table 2.1), based on Copernicus's data calculated from the mean sun (the mean sun is a point in space just off the actual or true sun). I am following Field (1988) by supplying percentage errors for the purpose of comparing the calculations from the mean sun in tables 2.1 and 2.2 to those from the true sun in table 2.3.

Kepler did not provide a separate table that included the moon, but he did tell us how to generate one (see Table 2.2):

> But if the lunar system is allocated to the thickness of the Earth's sphere, then if the inner surface of the Earth's sphere, also including the Moon is 1000 units, the outer surface of the sphere of Venus is 847 units in Copernicus. Also, the outer boundary of the Earth along with the Moon is 801 units if the lowest point of Mars is 1000. (*MC*, 157; *KGW* I, 48)

Kepler was greatly encouraged by his results:

> Hence you can realize how easily it would have been noticed, and how greatly unequal the numbers would have been, if this undertaking had been contrary to Nature, that is, if God himself at the Creation had not looked to these proportions. (*MC*, 157; *KGW* I, 50)

In chapters 15–7 Kepler compared the distances implied by the polyhedral hypothesis with the new distances calculated by Maestlin from the true sun, but did not end up with results that were much better (compare

TABLE 2.1
Polyhedral Distances without the Moon

				Polyhedral Distances	*Copernican Distances*	*Errors*
	Saturn		Jupiter	577	635	–9%
If lowest	Jupiter	*is 1,000* [a]	Mars	333	333	0%
point of	Mars	*highest*	Earth	795	757	+5%
	Earth	*point*	Venus	795	794	0%
	Venus	*should be*	Mercury	577	723	–20%
			or	707[b]		–2%

[a] That is, if the lowest point on the radius of the inner edge of the sphere of Saturn is 1,000, the highest point, or radius of the outer edge of the sphere of Jupiter is 577. Likewise for the other planets.

[b] For Mercury, the number 707 is derived not by the radius of the inner edge sphere inscribed inside the solid, as in the other spheres, but from the outer edge. Each sphere has certain thicknesses to account for the eccentric orbit. For a detailed discussion of Kepler's inscription and circumscription of spheres, see Field (1988, 35–44, 63–70).

TABLE 2.2
Polyhedral Distances with the Moon

				Polyhedral Distances	*Copernican Distances*	*Errors*
	Saturn		Jupiter	577	635	–9%
If lowest	Jupiter	*is 1,000*	Mars	333	333	0%
point of	Mars	*highest*	Earth	795	801	–1%
	Earth	*point*	Venus	795	847	–6%
	Venus	*should be*	Mercury	577	723	–20%
			or	707		–2%

tables 2.1, 2.2, and 2.3).[9] In Table 2.3, the Earth's mean distance is taken to be 1. Here Kepler is using sexagesimal units.[10]

It is intriguing that Kepler prevailed on Maestlin to calculate distances from the true rather than the mean sun. This important modification went against the whole history of astronomy (even Copernicus calculated from the mean rather than the true sun). Kepler was aware that such a move required justification, but he did not offer a detailed one:

> In case, friendly reader, I should offer you any occasion for rejecting the whole of this enterprise because of a trifling dispute, I must here mention to you something which I should like you to remember carefully: Copernicus's purpose was not to deal with cosmography, but with astronomy. That is, he is not much concerned whether there is a mistake relating to the true proportion of the

TABLE 2.3
Polyhedral Distances Compared with Distances from the True Sun

	Polyhedral Distances without Moon	*Polyhedral Distances with Moon*	*Distances from True Sun*	*Errors (Averaged) No Moon / Moon*
Saturn				
Aphelion	10 35′56″	11 18′16″	9 59′15″	+6% / +13%
Perihelion	8 51′8″	9 26′26″	8 20′30″	
Jupiter				
Aphelion	5 6′39″	5 27′2″	5 29′33″	−7% / −1%
Perihelion	4 39′8″	4 57′38″	4 59′58″	
Mars				
Aphelion	1 33′2″	1 39′13″	1 39′52″	−6% / 0
Perihelion	1 18′39″	1 23′52″	1 23′35″	
Earth				
Aphelion	1 2′30″	1 6′6″	1 2′30″	0% / 0%
Perihelion	0 57′30″	0 53′54″	0 57′30″	
Venus				
Aphelion	0 45′41″	0 42′50″	0 44′29″	+3% / −4%
Perihelion	0 42′55″	0 40′14″	0 41′47″	
Mercury				
Aphelion	0 30′21″	0 28′27″	0 29′19″	+2% / −4%
Perihelion	0 14′0″	0 13′7″	0 14′0″	

> spheres, but only with establishing from the observations the values which are best suited for deriving the motions of the planets and computing their positions, as far as possible. (*MC*, 159; *KGW* I, 50)[11]

Kepler's defense seems to be that cosmographical considerations support calculations based on the true rather than the mean sun. These considerations, given that he did not yet have the empirical evidence that he provided in the *Astronomia*, amounted to his Trinity analogy and a conception of physical causes such that they act both from and on bodies. Kepler hinted that geometrical points could not be the cause of motion,[12] but he did not articulate this as well here as he did in the *Astronomia*. Nonetheless, Maestlin was willing to undertake the project, which did not, for the purposes of the *Mysterium*, prove profitable, as the fit between theory and data was not improved.

At this point Kepler did not have the resources to determine whether the culprit was the calculated distances or the polyhedral hypothesis. He did not have access to Brahe's data. As a result, Kepler's next move was to provide theoretical rather than empirical support for the polyhedra, and for this purpose he turns in chapter 18 to a discussion of the relationship between theory and observation. Kepler pointed out that in addition to the possibility of observational error, observations were typically taken of planets in a particular position (acronychal),[13] and astronomical theory was still too insecure to permit confident extrapolation to other positions. For one thing, the eccentricities and their causes were insufficiently known to permit accurate calculations of the orbits. As a result, Kepler did not consider himself in a position to test the polyhedra, and so close agreement with the distances was all that could be hoped for at this point.

In chapter 20 of the *Mysterium* Kepler attempted to uncover a connection between the planet's period and its distance from the sun (a forerunner of the third law), and in chapter 22 he explored the connection between a planet's speed and its distance from the sun (a forerunner of the second law). As Stephenson notes, however, at the time of writing the *Mysterium* Kepler's grasp of mathematics and astronomy was relatively unsophisticated. Also, at this stage Kepler was looking for regularity in planetary motion, not offering a physical explanation for such regularities (Stephenson 1987, 8–9). This twin focus, on empirical regularities and archetypes, is reflected in Kepler's approach to natural philosophy. Only later in the *Apologia* and the *Astronomia* does Kepler emphasize the need for explanations of efficient causes.[14]

Kepler's Cosmology

As the sun is at the center of Kepler's physical universe, God is at the center of his archetypal universe. Kepler's God was a Platonic God, an aesthetician and a geometer, who created physical things to express aesthetically pleasing geometrical constructions. As a mind, God's creation of the universe was based on ideas, ideas of quantities—shape, number, and extension—in particular (*MC*, 95; *KGW* I, 24).

For Kepler, pure archetypes are distinct from geometrical archetypes. Pure archetypes are "in their simplest divine state of abstraction, even from quantities themselves, considered in their material aspect" (*MC*, 125; *KGW* VIII, 62).[15] By this Kepler seems to have meant that pure archetypes are ideas in the mind of God, or parts of God's essence, which are distinct from quantity:

> [T]he Creator conceived the Idea of the universe in his mind (we speak in human fashion, so that being men we may understand), and it is the Idea of that which is prior. . . . the only thing of which he could adopt the Idea for establishing the universe is his own essence . . . so that it [the universe] might become capable of accepting this Idea, he created quantity. (*MC*, 93–95; *KGW* I, 24)

To express the archetypes, or the divine essence, God created matter:

> [I]f we know the definition of matter, I think it will be fairly clear why God created matter and not any other thing in the beginning. I say that what God intended was quantity. To achieve it he needed everything which pertains to the essence of matter; and quantity is a form of matter, in virtue of its being matter, and the source of its definition. (*MC*, 93; *KGW* I, 23)

The archetypes are final causes in their divine state, formal causes in their material state, and God created physical forces such that the movements of bodies would express them (*MC*, 123; *KGW* I, 37–38). The result is a correspondence between final, formal, and efficient causes (*MC*, 93–95, 123; *KGW* I, 23–24, 37–38), a three-tiered universe, the geometrical linking the physical to the archetypal by serving as a spatial translation of the divine. Thus the levels represent, or mirror, each other in "architectonic harmony."[16]

The conception of different levels of reality was not novel, nor was the notion that a relationship of symbolic representation held between the levels. Plato's doctrine of participation is a forerunner of this view, and many Renaissance theorists followed Plato by postulating an architectonic harmony, where the corporeal is in harmony with the divine.[17] Unlike other Neoplatonists, however, Kepler treated the mirroring between levels of reality as direct, or faithful. He did not attempt to "transcend the world of the vulgar" in order to uncover the mystery of the universe. Rather, he *tested* the archetypal hypothesis against empirical data. This direct mirroring implies a commitment to two theses, that archetypal hypotheses are testable, and that empirical data are trustworthy indicators of the ultimate nature of reality.

Like others of his day, Kepler believed that the divine blueprint of the universe is reflected in human beings, that the human mind, though finite and fallible, is rational in the way God's is. For Kepler this meant that the mind has a priori powers that can be used to apprehend the archetypes. Once one knows the archetypal structure, one can deduce "the natural properties of the planets from immaterial things and mathematical figures" (*MC*, 123; *KGW* I, 37), a capacity possible only if the physical accurately reflects the archetypal.

Kepler's Approach to Natural Philosophy

Kepler did not write a treatise on the methodology of natural philosophy. The closest he came was the *Apologia*. Kepler tended to express his positions on issues like explanatory adequacy and testability only at junctures where he felt the need to justify certain claims. Even the *Apologia* was directed against a specific opponent, namely, Ursus. As a result, Kepler's approach must be extracted piecemeal from both his mention and use of methodological considerations. Throughout his works, however, he was explicit about one thing: that astronomy is capable of providing a *true* description of planetary motion. The picture that emerges throughout this study is that the conjunction of Kepler's metaphysics and epistemology provided him with a programmatic system of requirements for hypotheses to qualify as candidates for truth. By restricting the discussion to Kepler's philosophy as it emerges in the *Mysterium*, it will be clear that Kepler's philosophical acumen is appreciable even at this stage. Indeed, more so here than in his later works, Kepler addressed the issue of what kinds of hypotheses can count as explanatory candidates at all.

The Knowability of Nature

Kepler, in a letter to Maestlin, enthusiastically advocated the use of the a priori method, in contrast to Brahe's a posteriori method from observations to theory (October 3, 1595; *KGW* XIII, 33–40). Some comments in the *Mysterium* suggest that Kepler was advocating a kind of top-down method from the archetypes to the physical world. In chapter 11 Kepler proudly announced that he "deduced the natural properties of the planets from immaterial things and mathematical figures" (*MC*, 123; *KGW* I, 37). If Kepler meant that he did not need to confirm the archetypes empirically, this boast turns out to be empty. As already noted, Kepler uncovered the archetypes on the basis of rough empirical data and then checked them against more detailed observations.

On another, more charitable interpretation, Kepler believed that accessing the correct archetypes, by whatever means, allows one to deduce physical laws. Although Kepler was not an ontological reductionist, on this interpretation he held that the laws of one domain can be deduced from those of another. This reasoning seems more in line with Kepler's procedure in the *Mysterium* than with what he said about it. Later in a letter to David Fabricius, Kepler claimed not to develop archetypal hypotheses independently, but to construct them on the basis of observations (July 4, 1603; KGW XIV, 412).[18] As noted in the previous section, Kepler thought archetypal hypotheses must be both testable and tested before one can draw conclusions as to the aptness of the archetypal reasoning.

Nonetheless, the first interpretation has virtues to recommend it. Kepler believed that if one could know the archetypal structure independent of experience, one could then deduce all physical properties and laws and be confident in the correctness of these laws. Kepler was not alone in believing that knowledge in one domain could supply knowledge in another. In his study of Marsilio Ficino, for example, Kristeller notes that Ficino believed that because objects in the corporeal world represent a different, more divine realm, one can "know an existing thing not only through its direct attributes but also through its relation to that higher reality" (Kristeller [1943] 1964, 95). For Ficino, however, as lower-ranking substances in the hierarchy of being partake less of qualities like truth, empirical enquiry is inadequate for discovering truths about the divine realm.

Kepler's a priori method is designed to take human limitations into account, where the a priori and a posteriori are and must be used in a mutually supportive manner. Given our inherent limitations, the security of a hypothesis is incomplete until it is supported by arguments from all relevant domains. A priori reasoning by itself is hypothetical in the weakest sense. In introducing his comparison of polyhedral distances with those from Copernicus, Kepler wrote: "Let us now pass to the distances between the astronomical spheres and the geometrical derivations: if they do not agree, the whole of the preceding work has undoubtedly been a delusion" (*MC*, 149; *KGW* I, 43). A posteriori reasoning by itself amounts to a "lucky guess":

> For what could be said or imagined which would be more remarkable, or more convincing, than that what Copernicus established by observation, from the effects, *a posteriori*, by a lucky rather than a confident guess, like a blind man, leaning on a stick as he walks (as Rheticus himself used to say) and believed to be the case, all that, I say, is discovered to have been quite correctly established by reasoning derived *a priori*, from the causes, from the idea of the Creation? (*MC*, 97–99; *KGW* I, 26)

It might seem strange that Kepler referred to empirically adequate models as "lucky guesses," but one should bear in mind that he was well aware that observations alone are insufficient to decide between competing models. The choice between empirically adequate models, made without extra-empirical evidence, amounts to a lucky (or unlucky) guess.

The Evaluation of Hypotheses

There are several indicators of whether a hypothesis is causal. As we shall see, Kepler based these indicators on his conception of God. His conception of God, and the manifestations of God's essence in the three levels, provided "regulative principles" (a term used by Buchdahl 1972, 273), for empirical inquiry. Simplicity, unity, elegance, and rationality are part of God's essence, and hence are reflected, given the three-tiered universe, in physical structures. The implication is that a successful account of the universe must exhibit these principles.

To begin, archetypal reasoning is elegant and simple. Recall that Kepler based his ordering of geometrical figures in part on the grounds of simplicity, with the sphere as the most simple figure (*MC*, 101; *KGW* I, 27–28) and the cube the most simple of the regular solids (*MC*, 109; *KGW* I, 31). Because God's reasons for creating causal mechanisms are archetypal, physical causes will be simple as well. This grounds Kepler's assertion that nature is simple and "uses one cause for many effects" (*MC*, 77; *KGW* I, 16). As a result, a more fruitful hypothesis is more likely to have identified a causal relation. Two types of fruitfulness are of interest here, breadth of predictive range and breadth of explanatory range.

Kepler used the first type of fruitfulness to argue for a realist interpretation of astronomy. In doing so, he was responding to those critics of Copernicus who claimed that accurate observations can be derived from false hypotheses. Kepler argued that although false hypotheses can yield accurate predictions, such success is fortuitous and limited. The test for determining whether predictive accuracy is due to chance is applying the hypothesis to related phenomena that the hypothesis was not originally constructed to explain: "The conclusion from false premises is accidental, and the nature of the fallacy betrays itself as soon as it is applied to another related topic" (*MC*, 75; *KGW* I, 15). So for a hypothesis to be a contender for truth, it must be fruitful.[19]

Explanatory fruitfulness was also seen by Kepler as a significant virtue of hypotheses. Kepler argued that the Copernican hypothesis was superior in that it was able to answer questions that the Ptolemaic hypothesis rendered mysterious. For example, the Ptolemaic system cannot explain why the correction for the orbit of Mars is larger than that for Jupiter, and why that for Jupiter is larger than that for Saturn. The Copernican system explains these phenomena as a function of the distance from the moving Earth's orbit to the other orbits (*MC*, 81; *KGW* I, 18).

Kepler also supported his polyhedral hypothesis in a manner that evokes Lakatos's notion of fruitfulness, where progressive hypotheses explain novel facts:[20] "And thus we have in passing the reason why three planets move outside the Earth, two inside; and if this meets with less approval from the reader, let him reflect, *that this is a by-product, and not the main point*" (*MC*, 107; *KGW* I, 30–31, emphasis added). That a hypothesis can explain phenomena for which it was not originally intended is a striking success for that hypothesis.

For Kepler, however, the fruitfulness of a hypothesis is insufficient for its acceptance. Kepler added a proviso to the claim that a false hypothesis will be determined as such by broadening the range of application. A false hypothesis will not "betray itself" if

> you gratuitously allow the exponent of that argument to adopt an infinite number of other false propositions, and never in arguing forwards and backwards to reach consistency. That is not the case with someone who places the Sun at the center. For if you tell him to derive from the hypothesis . . . any of the phenom-

> ena which are observed in the heavens . . . he will have no difficulty with any point. . . . he will return with complete consistency to the same assumptions. (*MC*, 75; *KGW* I, 15)

Kepler seems here to have approximated the view that a hypothesis permitting too many ad hoc modifications is not sufficiently open to epistemic evaluation. The application of such a hypothesis to related phenomena will not reveal whether the hypothesis is false. For this reason the Copernican model was superior to the Ptolemaic, even though both were empirically adequate at the time. The Copernican model is able to constrain the number and time of the retrograde motions of the planets by making retrograde motion a function of the moving Earth. The Ptolemaic model, by contrast, saved the retrogressions by a series of ad hoc devices.[21]

Later, in the same paragraph as the preceding quotation, Kepler hinted at a theoretical virtue that would become very important in the *Astronomia*: a model that imposes sufficient constraints (one that does not permit many ad hoc modifications) is more testable. Kepler wrote that the Ptolemaic model, unlike the Copernican, is unable to explain why the different retrogressions "agree precisely, as they do" (*MC*, 77; *KGW* I, 15). Given the preceding content in the paragraph, what Kepler seems to have meant is that the same data provide more evidence for the Copernican model than for the Ptolemaic. The retrograde motions of the various planets are evidence for the same model in the Copernican system (indeed, they *measure* the same parameter, the Earth's motion), whereas in the Ptolemaic, the retrograde motions are treated individually (*MC*, 75–79; *KGW* I, 15–17).

Kepler did not restrict his methodological requirements to astronomy. He also required the same sort of rigor from his archetypal hypotheses. Consider Kepler's argument that abstract numbers cannot be the basis of a divine plan. Kepler's comments are cryptic, but his meaning can be reconstructed from various sources. In the *Mysterium*, when discussing the possibility of basing a rule of order for the distances on numerical relations, Kepler observed:[22]

> [O]n this basis whatever ratio I obtained, in whatever way, yet there would be no end to the calculation, no definite tally of the moving circles, either in the direction of the fixed stars, until they themselves were encountered, or at all in the direction of the Sun, because the division of the space remaining after Mercury in this ratio would continue to infinity. . . . For in discussing the foundation of the universe itself, one ought not to draw explanations from those numbers which have acquired some special (7) significance from things which follow after the creation of the universe. (*MC*, 65; *KGW* I, 10)

The (7) refers to a note of clarification, added in 1621:

> However, this particular circumstance, by which they are designated perfect numbers, considered on its own, does not have any significance either, that is,

> the circumstance that they are all numbers which are proper divisors of some particular number, and if added together are equal to the one of which they are divisors. Equality is indeed a fine thing; but this equality is an accidental property of those numbers if each one is considered separately, and does not relate in any way to their constitution, but the result of geometrical necessity which follows after they have been constituted. Nor does it impart to them the property of being more readily divisible, since it is a feature of this very divisibility, and is in a sense defined by it. (*MC*, 71; *KGW* VIII, 29)

In a letter to Maestlin, Kepler claimed that one should not look for archetypal relations in accidental properties, the relations between numbers being accidental to those numbers (October 3, 1595; KGW XIII, 34).

I take this to mean that when various numbers are connected to each other, the relationships between the numbers are not essential properties of the numbers themselves. The number 9, for example, can be related to the number 8 (8 + 1), or 7 (7 + 2), or 3 (3^2)—indeed, to an infinite number of numbers in an infinite number of ways per number. Nothing about the number 9 itself indicates which of these relationships we should pick. The choice is arbitrary. Geometrical figures, by contrast, *contain* relations. The cube, for example, contains the relation of equality of faces; one cannot construct a cube with irregular faces. In addition, general agreement that the sphere and the circle are the most perfect figures made it seem that one could rank-order polygons and polyhedra by degrees of similarity to the circle and sphere respectively (Kepler addressed this matter more rigorously in the *Harmonice*). It was not clear to Kepler how one could similarly rank-order numbers.

If we accept that one can rank-order geometrical figures, but not numbers, then an archetypal model based on geometry restricts the possibility for ad hoc modifications in a way that one based on number does not. If there is nothing special about numbers themselves that determines which are chosen and the order in which they are placed, then an infinite number of numerological archetypal accounts are possible. This idea was an anathema to Kepler, who believed that God would not create anything without a reason (*MC*, 123; *KGW* I, 38). Because God's reasons for constructing the universe were not arbitrary, the nature of the archetypes had to imply some types of constructions and not others. Kepler's God, then, is Platonic, not Pythagorean, for the appropriate model is supplied by geometry, not numerology. This is why Kepler was so pleased with his polyhedral hypothesis. There can be only five regular solids, which are themselves geometrically significant. That there are only five around which orbits can be inscribed determines the number of orbits.

These criteria are all expressions of the principles of unity and order. The universe, for Kepler, as a reflection of God, is characterized by the unification of diverse phenomena under hypotheses. But Kepler goes farther than asserting that the simpler theory is more likely to be true. His discussion of fruitfulness and model constraints is an account of why hypotheses that are both simple and broad in scope are more epistemically valuable. The world was created to be knowable. Kepler viewed properly constructed and tested hypotheses as indicators of causal mechanisms, suggesting that since the Copernican hypothesis met the requirements of construction and testing, it revealed the nature of reality: "On all these points, as a magnificent order is shown by Copernicus, the cause must necessarily be found in it" (*MC*, 77; *KGW* I, 15). Kepler corrected and emphasized the relationship between proper hypotheses and the uncovering of causes in 1621: "Furthermore I was nodding when I composed [the preceding sentence] . . . for I meant to say, that as a magnificent order is shown by Copernicus, as there is in the relationship between a cause and its effects, the true cause of the retrogressions must necessarily be this very fact" (*MC*, 87; *KGW* VIII, 40). The suggestion, I take it, is that given the nature of the order of the physical universe, a theory that expresses this order should be seen as an accurate representation. Given the architectonic harmony, the rational and the physical correspond.

In addition to grounding internal methodological commitments, Kepler's three-tiered universe implied that confidence in a given hypothesis would be increased if it appeared prima facie capable of fitting with hypotheses from other domains. It is easy to see why. Any account that is an unlikely candidate for harmonizing with those from other domains is an unlikely candidate for being an accurate hypothesis in its own. Recall Kepler's comparison of the polyhedra and musical harmonies. As noted before, Kepler's aim was to show that the pattern exhibited by the polyhedra occurs elsewhere in nature, and hence is an archetypal candidate. Although Kepler did not ground his astronomical account in physics in the *Mysterium* the way he did in the *Astronomia*, or the way he argued one ought to in the *Apologia*, his conception of recurring archetypal patterns in the various disciplines was an important conceptual precursor to his construction of physical astronomy. A hypothesis with converging support from other domains is more secure than one without. Kepler considered it highly unlikely that he could get agreement from such diverse sources if the hypothesis had been false (*MC*, 97, 99; *KGW* I, 26). Again, we see the idea that the convergence of evidence toward a single conclusion makes it very unlikely (indeed, a remarkable coincidence) for that conclusion to be false.

One wonders whether at this point Kepler clearly distinguished between agreement from diverse empirical sources and agreement from diverse theoretical sources. It seems at first glance that he did. His analysis of the internal unification of the Copernican system appears in a different chapter from his claim that the coincidence would be too great if the agreement between the astronomical and archetypal disciplines were spurious (chapters 1 and 2 of the *Mysterium* respectively). However, Jardine (1984, 223–24) has argued that this division reflects not an awareness of the distinction between empirical and theoretical unity, but a concern with the distinction between that which was traditionally considered part of a given discipline and that which was not obviously relevant to that discipline.

In any case, the emerging picture of Kepler's requirements for an adequate causal hypothesis is quite complex and sophisticated. What I hope is now clear is that Kepler already had an account of empirical and theoretical adequacy that was considerably more interesting than the requirement that one save the phenomena any way one can (the two types of fruitfulness, model constraints, and candidacy for fitting an archetypal hypothesis), and that these requirements applied to both astronomical and archetypal hypotheses (later in the *Astronomia* Kepler applied these criteria to physical hypotheses as well). Meeting these requirements simultaneously establishes the security and the acceptable form of a hypothesis. What I hope is also clear is that Kepler's archetypes played a positive role in motivating these methodological commitments.

In addition to motivating his methodological approach, Kepler's philosophical commitments provided resources for dealing with skepticism, as we shall see in more detail in the next chapter. One of the major skeptical challenges to astronomy was the existence of observationally equivalent, incompatible astronomical hypotheses. Kepler's conception of the world as a mirroring of ontologically distinct but nomologically reducible domains gave license to the use of considerations from other domains in adjudicating between astronomical hypotheses.

3

Kepler's *Apologia*: An Early Modern Treatise on Realism

THE *APOLOGIA PRO TYCHONE CONTRA URSUM*, Kepler's most epistemologically focused work, was written around 1600 but not published until 1858. Jardine speculates that had it been published in Kepler's lifetime, and had it received due attention in the interim, it would now "be a classic on a par with such seminal reflections on the nature of human inquiry as the *Novum organum* and the *Discourse on Method*" (Jardine 1984, 5).[1] Instead, like many of Kepler's works, it has been studied only by a few. What is rather striking about the *Apologia* is that Kepler anticipated and provided sophisticated approaches to certain problems facing modern-day scientific realists, to wit, the problems of empirical equivalence, and the related problem that empirical success is not necessarily an indication of truth.[2] Another respect in which the *Apologia* should be of interest to philosophers of science is that Kepler's grappling with philosophical issues engendered his novel approach to, and subsequent discoveries in, physical astronomy.

Although the *Apologia* reads as though it could have been written in the twentieth century, the roots of Kepler's realism are to be found in his Neoplatonic cosmology.[3] My purpose here is to trace Kepler's response to the skeptics and to show how his response was motivated and justified by his cosmology.[4] Of particular interest here is that whereas the focus in the *Mysterium* was primarily on formal and final causes of astronomical phenomena, in the *Apologia* we find much more attention paid to efficient causes.

The *Apologia*, though a philosophical treatise, was directed at a specific opponent, namely the imperial mathematician Nichol Baer (or Ursus), in defense of Tycho Brahe in his priority dispute over the invention of the Tychonic system.[5] Kepler was obliged to write the *Apologia* in order to gain access to Brahe's remarkable observations. Kepler, embarrassed by the very idea of a priority dispute, compromised by writing a philosophical treatise on the historical and current status of astronomical hypotheses. Although Kepler did not directly address the charge of plagiarism, the *Apologia* did function as a defense against Ursus's claims that the geoheliocentric system could be found in the works of the ancients and that, in any case, it was a simple matter to construct a variety of empirically adequate hypotheses. In response to these attacks on Brahe's claims to originality and to having based his geoheliocentric system on observations, Kepler argued that Ursus had misinterpreted the ancients and misconstrued the proper process of constructing and justifying astronomical hypotheses (Jardine 1984, 3–5).

Given the purpose for which the *Apologia* was written, it cannot be taken as a full articulation of Kepler's approach to natural philosophy. In particular, because Kepler's use of the polyhedral model supported the Copernican over the Brahean system, it should not be surprising that the archetypes were not given a place of prominence in the *Apologia*. Nonetheless, the *Apologia* provides valuable resources on Kepler's notions of the proper form for astronomical hypotheses and the proper means of testing them. Jardine argues that Kepler's epistemological attitude toward astronomy was new and proposes "that if any one work can be taken to mark the birth of history and philosophy of science as a distinctive mode of reflection on the status of natural science it is Kepler's *Apologia*" (Jardine 1984, 2). As already noted, this innovation, I believe, was motivated by Kepler's cosmology.

Ursus, like many of his day, viewed astronomical hypotheses as mere predictive devices. Jardine (1984, 232–41) argues that this epistemological attitude was taken to avoid the conflict between physics and empirically adequate astronomical hypotheses. One of the arguments for the skeptical position was based on the availability of various different hypotheses that yielded the same predictions. Given that incompatible extant hypotheses yielded the same predictions, predictive accuracy could not be seen as guaranteeing truth.

Kepler responded to the problem of observational equivalence on three fronts: hypothesis testing, the nature and interpretation of hypotheses, and the use of external considerations. His conclusion is simple and effective; hypotheses properly formulated and tested are not evidentially equivalent, or else they are the same hypothesis. I address the three components here in reverse order since the latter two were already discussed in chapter 2.

Extra-Astronomical Considerations

Kepler suggested that observationally equivalent hypotheses could be differentiated, if considerations from other disciplines were adduced:[6] "And though some disparate astronomical hypotheses may provide exactly the same result in astronomy, as Rothmann claimed in his letters to Lord Tycho of his own mutation of the Copernican system, nevertheless there is often a difference between the conclusions because of some physical consideration" (*Apologia*, 141; *KGW* XX.1, 21). Physical considerations, for Kepler, included physical forces and archetypal constraints. In this sense, one of the physical differences between the Copernican and Ptolemaic hypotheses is that the former but not the latter allow for Kepler's polyhedral hypothesis.

The proposal to include physics in astronomy was not a radical one. As Duhem ([1908], 1969 113–17) disapprovingly notes, the history of astronomy is full of examples of astronomical theories constrained by physics. Aristotle, for example, required on physical grounds that the planets circled the Earth concentrically. Kepler, of course, approved, arguing that many of the advances made in the history of astronomy were the result of the inclusion of such physical concerns (*Apologia*, 177; *KGW* XX.1, 43).

Kepler's attitude toward the use of physics in astronomy is somewhat surprising, given that, at the time, no empirically adequate astronomical hypotheses were consistent with the available physics. The details of his own physical astronomy had not yet been worked out. By the time he had written the *Apologia*, he had uncovered what to him was a striking fit between an astronomical hypothesis (the Copernican) and an archetypal account. He was also aware of William Gilbert's work on magnets and thought it filled a gap in the arguments for the Copernican system presented in the *Mysterium* (*Apologia*, 146; *KGW* XX.1, 24). But even without having developed his new astronomy in detail, these preliminary results seemed to Kepler sufficiently promising to justify his faith that a consistent physical astronomy was forthcoming.

This surprising attitude (for its time) was justified by the belief that true theories and the archetypes mirror each other. Becasue there are only a few archetypal patterns, but many different kinds of objects in the world, one should expect to find the same archetypal pattern in different disciplines. Hence, developments in one discipline *should* influence other disciplines, a much stronger norm than merely permitting physical considerations to constrain astronomy. Kepler's goal was to merge disciplines to create a consistent, unified picture of the universe. Because in his day it was not considered methodologically sound to require importation of extradisciplinary considerations,[7] Kepler needed to justify the practice of broadening evidential grounds; in this case the justification to which he appealed was the architectonic harmony of the universe. Given the historical significance of Kepler's use of physics in astronomy, one cannot overemphasize the positive role the archetypes played in providing a link between disciplines, thereby enabling the convergence of theoretical support from various domains.

The Nature of Astronomical Hypotheses

A crucial aspect of Kepler's realism is revealed in his distinction between geometrical and astronomical hypotheses. An astronomical hypothesis is based on the actual orbit of the planet, whereas a geometrical hypothesis is a means of constructing the orbit (*Apologia*, 143, 153; *KGW* XX.1, 22–23). This distinction had two functions: it reduced the number of hypotheses under consideration, and it was part of Kepler's diagnosis of what he saw as the fundamental confusion underlying the skeptical claim that one cannot adjudicate between astronomical hypotheses.

Kepler addressed what he called "perhaps the best-known example" (*Apologia*, 142; *KGW* XX.1, 22) of observationally equivalent hypotheses adduced by the skeptic, the concentric with an epicycle versus the eccentric. He concluded, in contrast to the skeptic, that the two hypotheses are not competing because they represent the same planetary motion—"that although the motion of the planet remains steady, it spends a longer time in one half of the circle" (*Apologia*, 143; *KGW* XX.1, 22). The difference between the two is in the geometrical method of constructing the orbit, not the orbit itself. The difference between the two, then, is one that makes no difference to astronomy.

Kepler identified the conflation of astronomical and geometrical hypotheses as the root of the skeptic's argument that the consequences of a hypothesis, if true, do nothing to increase one's confidence in that hypothesis. Competing astronomical hypotheses, in contrast to geometrical ones, will differ in their empirical consequences since the motion and path of the planet itself will be portrayed differently.

One wonders why the distinction between geometrical and astronomical hypotheses was not obvious. Indeed, Kepler thought the skeptic missed the obvious: "No one is to be taken seriously who does not acknowledge and grasp this diversity in hypotheses. Indeed, Osiander makes a mockery of Copernicus, or rather of his readers, by his equivocation, transferring those things that are true of geometrical hypotheses to astronomical hypotheses, even though the natures of the two sorts are very different" (*Apologia*, 153–54; *KGW* XX.1, 28). Geometrical hypotheses are used simply to predict planetary positions, while astronomical hypotheses represent planetary motion. To be fair to Kepler's opponents, before Brahe it was reasonable to blur the distinction. When using the Eudoxian method, the circles deployed to construct the orbit represented the phenomena of principle interest—that of the solid spheres. The same holds for the Ptolemaic system if one follows Peurbach's physical interpretation. The distinction between astronomical and geometrical hypotheses only became of interest after Brahe's argument that the solid spheres do not exist. Only then did the orbit of the planet take precedence over the spheres that the circles represented. Only then could different geometrical constructions be seen as representing the same phenomenon. Prior to that they represented different physical systems with the same observational consequences.

Kepler, however, saw the source of the conflation as a confusion of the more basic distinction between appearance and reality. In astronomy, in particular, the distinction seemed necessary. The irregular apparent motions of the planets seemed tractable to no rule if taken at face value, which is in conflict with belief in the perfection of the heavens. Kepler mocked Patricius, whom he interpreted as claiming that the planets move exactly as they appear to: "For where, walking through the fields, he encounters hedges and things near to his path, he would believe, on the testimony of the sense of sight, that distant mountains are really following him" (*Apologia*, 155; *KGW* XX.1, 29).[8] The other extreme, viewing the apparent as entirely distinct from the real, was motivated and compounded by the radical segregation of the heavens and Earth, each with a dramatically different metaphysical nature. Iacopo Zabarella, for example, conceived of heavenly "matter" as having "no potential for existence, but only for location."[9] Small wonder that some viewed the heavens as inaccessible to human inquiry (e.g., Giovanni Gioviano Pontano [1426–1503]).

Kepler occupied the middle ground between these two extremes. He did not deny that observation could contribute to knowledge acquisition. Indeed, he required such input, indicating in a rather cryptic passage that distinguishing the various tasks of astronomy will illuminate the relationship between appearance and reality:

> To set down in books the apparent paths of the planets and the record of their motions is especially the task of the practical and mechanical part of astronomy; to discover their true and genuine paths is (despite Osiander's and Patricius' futile protests) the task of contemplative astronomy; while to say by what circles and lines correct images of those true motions may be depicted on paper is the concern of the inferior tribunal of geometers. He who has learned to distinguish these things will easily disengage himself from the deluded seekers after abstract forms who quite heedlessly despise matter (the one and only thing after God), and from their importunate sophisms. (*Apologia*, 156; *KGW* XX.1, 30)

The mistake implicated here is the assumption that the only alternative to recording apparent motion is to determine a priori the appropriate geometrical means of constructing planetary motion (a mistake made by "the deluded seekers after . . . forms"). Kepler saw this as a false dichotomy, pointing to the possibility that there is a relationship between apparent motion and the actual orbit. The apparent motions are caused by the actual orbit and, in turn, yield information about it.

Notice that the distinction between appearance and reality does not lead to regarding observation as misleading and unnecessary for exploring the nature of reality. What it leads to instead is a recognition of the distinct roles of observation and reason, which imply specific *relationships* between the distinct tasks of astronomy. These tasks and their relationships are designed to take into account the process of discovery available to the human mind:

> In all acquisition of knowledge it happens that, starting out from those things which impinge on the senses, we are carried by the operation of the mind to higher things which cannot be grasped by any sharpness of the senses. The same thing happens also in the business of astronomy, in which we first of all perceive with our eyes the various positions of the planets at different times, and reasoning then imposes itself on these observations and leads the mind to recognition of the form of the universe. And the portrayal of this form of the universe thus derived from observations is afterwards called "[a body of] astronomical hypotheses." (*Apologia*, 144; *KGW* XX.1, 23).

Recognizing the relationships between observation and theory, and between appearance and reality, leads to a denial of skepticism. Given that, for Kepler, the material is a reflection of the rational, observations are amenable to being ordered by reason, and hence both the task of prediction and that of explanation are within the province of the astronomer.

Notice the difference between the preceding quotation and Kepler's apparent boast in the *Mysterium* to have developed astronomical hypotheses a priori. As already discussed in chapter 2, Kepler should not be taken as asserting that one can ignore observations in uncovering the true nature of the heavens. Here Kepler is explicit about the role of observations as precedents to and guides of theorizing:

> Next, he [Ursus] says that hypotheses are contrived for the purpose of observing the celestial motions. This view is diametrically opposed to the truth. We can observe the celestial motions even if we hold absolutely no opinion about any dispositions of the heavens. The fact is that observation of the celestial motions guides astronomers to the formation of hypotheses in the right way, and not the other way round. (*Apologia*, 144; *KGW* XX.1, 23)

The relationship between hypotheses and observations is that observations lead to and constrain hypotheses in the process of discovery. The theory is arrived at by the rational ordering of data. Thereafter one can infer observations from the theory. Kepler's advocacy of the a priori method is not a denial of the importance of empirical data but an emphasis of the role reason plays in uncovering the order underlying the data.

Kepler, however, was well aware that one could impose a false order on data. He thought any such order should be evaluated both against the proper testing of hypotheses and by its fit to considerations traditionally considered external to the discipline.

The Testing of Astronomical Hypotheses

In his analysis of the proper testing of astronomical hypotheses, Kepler responded to the claim that one can elicit true conclusions from false premises by arguing that when this happens it is by chance alone, and proper testing all but eliminates this possibility:

> And just as in the proverb liars are cautioned to remember what they have said, so here false hypotheses, which together yield the truth once by chance, do not in the course of a demonstration in which they have been combined with many others retain this habit of yielding the truth, but betray themselves. Thus in the end it happens that because of the linking of syllogisms in demonstrations, given one mistake an infinite number follow. (*Apologia*, 140; *KGW* XX.1, 21)

Kepler's discussion here is brief, but the moral is, by and large, the same as that in the *Mysterium*. As we saw earlier, the notion that false hypotheses will betray themselves was justified by what Kepler required of hypotheses generally: they must be applied to a variety of related phenomena, and there must be a restriction on the use of ad hoc propositions to save them from disconfirming evidence (*MC*, 75; *KGW* I, 15). The "linking of syllogisms" provides a method of checking for errors. Consequently, a hypothesis that accounts for a wide range of phenomena is preferable to one that applies to a narrow range; likewise, one that requires fewer ad hoc propositions.

There is a slight difference between how Kepler discussed the manner in which false hypotheses betray themselves in the *Mysterium* and in the *Apologia*. In the *Mysterium* Kepler argued that a false hypothesis will betray itself when applied to diverse *data*. Kepler spoke of being able to derive from true hypotheses "any of the phenomena which are actually observed in the heavens" (*MC*, 75; *KGW* I, 15). In the *Apologia*, he made the more general claim that a false hypothesis will betray itself when linked with other *hypotheses* during demonstrations. This requirement entails internal consistency more than it does consistency with data.[10] Kepler's reference to the *Mysterium*, however, suggests that he was speaking of both theoretical and empirical support—that if *all* the pieces fit together, then it is highly unlikely that any one part is false. This interpretation supports Jardine's (1984, 223–24) claim that the distinction between empirical and theoretical evidence had not yet been precisely drawn.

This emphasis on the internal relations of a theory arises directly from Kepler's aesthetic conception of the universe as unfolding from a small set of simple principles. The variety and apparent disorder that one finds in the world are really an expression of a few aesthetic generative principles. Consequently, the generation of the complex from the simple should be reflected in and a feature of a good theory.

Kepler's Improved Notion of Testability

Kepler's *Apologia* is of historical interest for the scientific realist for at least two main reasons. First, one can *read into* the *Apologia* an anticipation of several issues in the current realism-antirealism debate, in particular the problems of empirical equivalence and of the empirical success of false theories. Part of Kepler's approach to these problems was to widen the notion of evidence to include theoretical plausibility, an approach discussed by Boyd ([1983] 1991). Kepler's commitment to realism and his concern to answer the skeptic also led to the development of a more rigorous notion of testability. This notion was crucial to Kepler's radical innovations in astronomy, a point made clear in Kepler's conception of the proper testing of astronomical hypotheses in the *Mysterium* and the *Apologia*. Kepler required of hypotheses that they be broad in scope, both predictively and explanatorily fruitful, and of theories he required that their structure permit the linking of syllogisms.

As already noted in chapter 2, the requirements of breadth and fruitfulness serve to separate the true from the false in two ways. First, a greater variety of applications increases the opportunities for testing a hypothesis, as one is simply more likely to uncover recalcitrant data. Second, the application of a hypothesis to data not considered in its original construction discounts the possibility that a test is rigged.

Kepler's requirement of fruitfulness thus turns the limitations of the human mind into a testing advantage. If we were omniscient, then we could construct a hypothesis to account for all of the relevant data. But since we are not, we should expect that some of the data predicted by a hypothesis will surprise us. Kepler argued that our limited knowledge is not so much a liability but a testing opportunity. (Of course, we would not need to construct hypotheses if we really were omniscient.)

Recall that Kepler required a restriction on ad hoc modifications; otherwise the breadth and fruitfulness of a system of hypotheses will not be sufficient to warrant its acceptance.

> The conclusion from false premises is accidental, and the nature of the fallacy betrays itself as soon as it is applied to another related topic—unless you gratui-

> tously allow the exponent of that argument to adopt an infinite number of other false propositions, and never arguing forwards and backwards to reach consistency. That is not the case with someone who places the Sun at the center. For if you tell him to derive from the hypothesis . . . any of the phenomena observed in the heavens, to argue backwards, to argue forwards, to infer from one motion to another . . . he will have no difficulty with any point. (*MC*, 75; *KGW* I, 15)

In the *Apologia*, he spoke of the linking of syllogisms as a method for revealing errors. If the theory itself entails these links, all the better. A theory whose parts "hang together" is more testable than one with a number of independent hypotheses, because determining the values of a parameter for one hypothesis will have implications for the expected values of other parameters.[11] This restriction is *internal to the theory* on ad hoc modifications, rather than imposed from the outside. It is, if you will, a non–ad hoc restriction on ad hoc modifications.

The linking of syllogisms implied in the Copernican model, but not the Ptolemaic, is what makes the former more testable, even though they both save the same phenomena. Nothing in the Ptolemaic system rules out saving the phenomena by a series of ad hoc devices (e.g., by changing the size and number of the epicycles). The core constraints in the Ptolemaic system are that the Earth is stationary, and all motions around it be resolved into circles. These constraints do not determine the specific retrograde motions or even entail that such motion occurs. More important, changing the model for the retrograde motion of one planet neither affects the models for the other planets nor displaces the Earth. Under the Copernican system, by contrast, the retrograde motions are the direct result of the Earth's orbit. The various retrograde motions are unified under and linked with the hypothesis that the Earth moves.

As noted, Kepler's conception of the testability of hypotheses is related to his conception of divine aesthetics. The world is aesthetically beautiful, and so beneath the chaotic and diverse appearances, there is a simple order, a pattern to be discovered. It is to be discovered by showing the links between diverse phenomena, links that explain how the chaos of appearances is generated by a harmonious and simple pattern. The epistemological warrant here comes, however, not from the aesthetic features but from the testability of a theory. A brief look at the harmonic features of the Ptolemaic system should clarify this point.

To save the phenomena under the Ptolemaic system, the epicyclic motion of the planets is harmoniously ordered with respect to the position of the sun in the order of orbits. The inferior planets (inside the sun's orbit) never stray far from the sun, and go into retrograde motion only when they pass the sun. One way to ensure that the model produces these results is to place the center of the epicycle on the line between the Earth and the sun

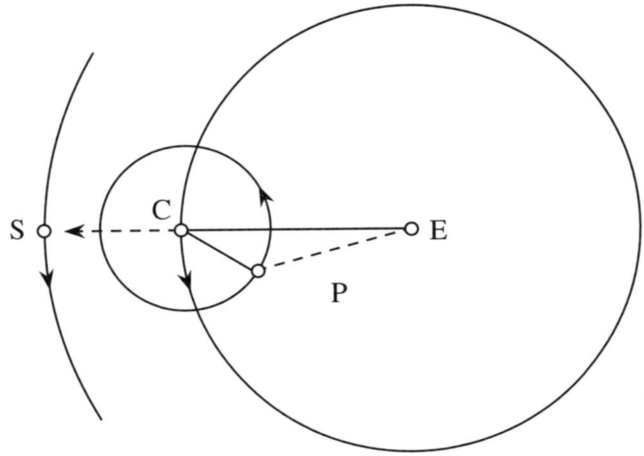

Figure 3.1. The Inferior Planets

(figure 3.1, where the sun is at S, the center of the epicycle at C, and the Earth at E). The retrograde motions of the superior planets only occur when the planet is in opposition to the sun, that is, when the Earth is on a line between the sun and the planet. One way of modeling this is to have the line between the center of the epicycle and the planet always parallel to the line from the Earth to the sun (figure 3.2, where the sun is at S, the Earth at E, the center of the epicycle at C, and the planet at P). In addition, to get the size and number of retrograde motions right, the size of the epicycle decreases the further the orbit is from the sun.

These features are aesthetically pleasing, one might argue, but they do not follow from the central conception of the Ptolemaic system, and thus are not the result of how the theory "hangs together." Under the Ptolemaic system, although there is an appealing order in the construction of the retrograde motions, they do not provide information about the solar orbit, because that does not explain the retrograde motions. Prior to noticing the harmony between the solar orbit and the retrograde motions, there is no reason to expect to find such a harmony, since the Ptolemaic system makes no such predictions. The data, then, do not provide evidential support for the construction of the solar orbit, but only for the construction of the individual models of retrograde motion. Thus, while the harmony of hypotheses provides propaganda for the Ptolemaic system, the agreement of hypotheses provides *evidence* for the Copernican, because this agreement was predicted by it. By contrast, the Ptolemaic system does "not give the reasons for the . . . retrogressions, and why they agree precisely, as they do" (*MC*, 75–77; *KGW* I, 15). Also, because the hypotheses are linked under the Copernican, determining the values for one (say, the retrogres-

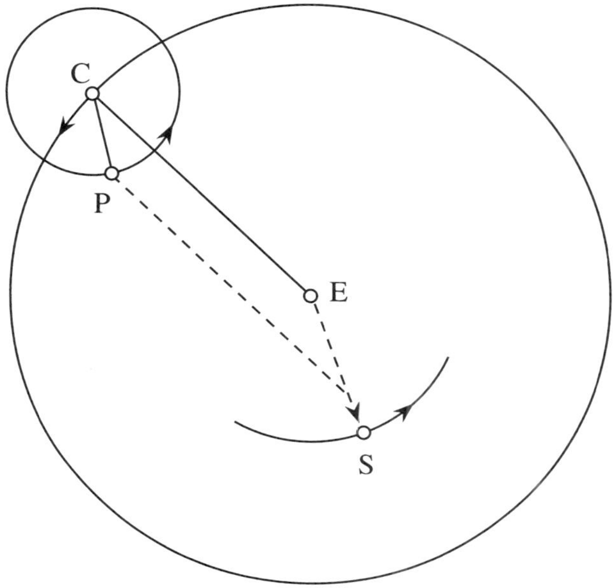

Figure 3.2. The Superior Planets

sions of Saturn) will influence values for another (say, the retrogressions of Mars), since they are both caused by the same thing, the Earth's motion. Thus one can generate more evidence from a fixed data set. The data remain the same, but more predictions are generated from them.

The aesthetic appeal of the Ptolemaic system is not the same as that of the Copernican and, as Kepler argued, that of the Copernican is preferable on evidential grounds. Kepler's link between aesthetics and testability allows the same data to provide more evidence for the Copernican system. This link is part of Kepler's answer to the problem of observational equivalence. While the Ptolemaic and the Copernican models may have been observationally equivalent, they were not evidentially equivalent.

Concluding Remarks: The Archetypes

Kepler's focus in the *Apologia* on defending the status of astronomical hypotheses leaves little room for a discussion of his archetypal universe. Indeed, given the purpose for which the *Apologia* was written, the inclusion of archetypes that supported the Copernican system over that of Brahe would have been impolitic. One can see, however, the relevance of Kepler's archetypes to his arguments against skepticism. Jardine (1979, 167) comments that Kepler did not justify his method as truth-linked. Although he did not explicitly do so in the *Apologia*, his answer to Jardine's charge is the archetypal vision (though this would hardly please the modern realist).

The use of extra-astronomical considerations (specifically, physics) in astronomy is legitimized by the notion that various objects and events studied by different disciplines are reflections of the same archetypes. Indeed, given Kepler's cosmology, converging support from various disciplines is required for the security of a hypothesis.

His conception of the proper testing of hypotheses rests on his vision of the universe as both orderly and unified. False hypotheses will betray themselves precisely because they can legitimately be applied to diverse phenomena. The view of the universe as explainable by a small set of principles gives one the means to determine whether a given hypothesis is false by drawing evidence from a variety of sources.

Kepler's view of the relationship between appearance and reality is predicated on the idea that material substance is obedient to a divine rational plan and, indeed, was created to instantiate the archetypes (*MC*, 125; *KGW* VIII, 62). So while there is a distinction between appearance and reality, there is also a systematic relationship between them that can be explored and exploited by observing appearances in tandem with considering the causes a priori. Each method is buttressed by the other, by requiring that the conclusions from both cohere.

On the basis of his arguments in the *Mysterium* and *Apologia*, Kepler was able to draw a distinction between empirical adequacy and truth. As we shall see in the next chapter, this distinction, with its attendant methodological and cosmological commitments, helped Kepler in his struggle to uncover the true orbit of Mars.

4

Kepler's Archetypes and the *Astronomia nova*

THE *ASTRONOMIA NOVA*, published in 1609, is probably Kepler's greatest contribution to astronomy, containing the first articulation of what are known today as Kepler's first two planetary laws. Having access to Brahe's data at last, Kepler advanced his "war on Mars." But, as Stephenson observes, even Brahe's data was insufficient to decide between competing theories: "Kepler's mathematical ingenuity was such that he could invent alternative theories, each good enough to "save" even the excellent observations of Tycho Brahe. These observations, therefore, did not provide enough guidance for him to reach his final solution to the problem of planetary motion" (Stephenson 1987, 2). To resolve this problem, Kepler applied the method recommended in the *Apologia* of grounding astronomy in physics. As its full title announces, the *Astronomia* is an investigation of the motion of Mars through considerations of its physical causes.[1] I argue here that Kepler was also guided by archetypal considerations, and that when he spoke of grounding astronomy in physics, he meant this in the broad sense used in the *Apologia*, where physics includes cosmology.

Kepler's archetypes seldom make an explicit appearance in the *Astronomia*. In Kepler's arguments to the ellipse they rarely play the roles assigned to them by the *Mysterium* and *Apologia*, as part of an a priori inference to the same phenomena, or as adjudicating between competing astronomical hypotheses.[2] Kepler's belief in an archetypally structured universe, I believe, nonetheless guided his physical exploration of the Martian orbit by narrowing the scope of hypotheses under consideration. Given the historical context, this approach is of great significance, for Kepler was the first to combine physics and astronomy in the manner he did, and he faced an open field of possible theory types. Kepler also needed to develop a new physics, because traditional Aristotelian physics was inappropriate to the task. His archetypes allowed him to rule out as implausible a number of physical investigations. Moreover, when Kepler failed to provide an account of the efficient causes of planetary motion, his archetypal commitments preserved his confidence that he was, nevertheless, on the right track.

After discussing Kepler's epistemological attitude toward hypotheses in physical astronomy, I trace the use of archetypal reasoning in his arguments to the ellipse, which involved establishing the true sun as the basis for calculations, determining the physical consequences of viewing the sun as emitting the primary motive force, and uncovering the true path of the Martian orbit. I follow Kepler's argument to the ellipse as he presented it, discussing some of the models he rejected along the way.[3]

The Status of Hypotheses in Physical Astronomy

Kepler believed that one could, with reasonable confidence, uncover truths about the nature of the universe. As noted in chapter 3, Kepler was well aware that empirical adequacy was insufficient to ground confidence in a given hypothesis. His solution was to obtain converging theoretical support from various disciplines, and to offer a number of types of arguments within each discipline to render each hypothesis individually plausible. One should note that these are two very different types of unification. Interdisciplinary unification involves the agreement of results—in this case, consistency of results from physics, astronomy, and cosmology. This sort of unification did not entail the conflation of disciplines. As we shall see, Kepler held that while their results must agree, the disciplines still have distinct objects of study and are therefore distinct themselves. The second kind of unification, the use of different arguments to the same conclusion within a single discipline, involves obtaining agreement from diverse phenomena, using different phenomena to measure the same parameter (e.g., showing that both the apsidal and nodal lines pass through the sun). Even so, Kepler was careful in drawing conclusions, for despite his confidence in the *Astronomia*'s end results, he pointed out that hypotheses could only be made plausible, not certain (*AN*, 47; *KGW* III, 19).

Part of the reason for Kepler's lack of certainty was his awareness that the physical intuitions he appealed to were novel and speculative. He also considered physics to be speculative by its very nature, because it is not based entirely on observations:

> I mingle the probable with the necessary and draw a plausible conclusion from the mixture. For since I have mingled celestial physics with astronomy in this work, no one should be surprised at a certain amount of conjecture. This is the nature of physics, of medicine, and of all the sciences which make use of other axioms besides the most certain evidence of the eyes. (*AN*, 47; *KGW* III, 19)[4]

This quotation is rather illuminating. Kepler considered physics to be speculative, yet believed that grounding astronomy in physics would adjudicate between competing astronomical hypotheses. This position highlights the difference between taking astronomical hypotheses to be secured by physics itself, and taking them to be secured by converging support from diverse areas, secure or otherwise. Kepler viewed the mingling of astronomy and physics as increasing security for the latter reason, clearly.

The Replacement of the Mean by the True Sun

One of the first things Kepler did in the *Astronomia* (pt. I, chs. 1–6) was to argue that both physical and astronomical models should be constructed around the true rather than the mean sun. The latter is an empty point that, under the Copernican system, is the center of the Earth's orbit. Gingerich comments that "The idea is so important that we should perhaps call it Kepler's zeroth law. It is a sign of Kepler's genius that he so quickly recognized this as the crucial first move in the reform of astronomy" (Gingerich 1975b, 264).

Kepler argued for his "zeroth law" at various points throughout the *Astronomia*, with each argument depending on previous results.[5] Kepler's "modest" goal in part I was to render the use of the true sun plausible, to argue that a different orbit is derived if the true sun is used instead of the mean sun, and to point out the observational differences between the two models (Kepler did this for the Copernican, Brahean, and Ptolemaic models).

The empirical differences turn out to be quite significant. Figure 4.1 depicts the differences in apsidal line and orbit when the true sun and the mean sun are used in the Copernican system.[6] If the mean sun is used as the basis for constructing the model, the equant point fixed at E, and observations taken as if from the mean sun at B, then the apsidal line will be placed through points B and K. This arrangement is so, Kepler argued, because, though the planet is actually moving slowest at D, it will appear to be slowest at K, since K is further than D from B. The point furthest from B is N, but Kepler pointed out that while motion appears slower the further away it is, the planet is actually moving slowest at D, and these two effects counteract each other, making the motion appear slowest at K (*AN*, 145; *KGW* III, 80). If, by contrast, the true sun A is used, with E as the equant point, then the point at which the planet appears slowest is at D, placing the apsidal line through points A and D. As a result, the apsidal line is placed differently in the two models. In figure 4.1, the dotted line represents the orbit and apsidal line of the mean sun model, and the solid line represents the true sun model. Not only is the orbit shifted over, but the time taken to traverse equal arcs would be asymmetrical to the apsidal line in one of the models. Kepler was confident that this asymmetry would be found in the mean sun model, though he did not offer empirical support

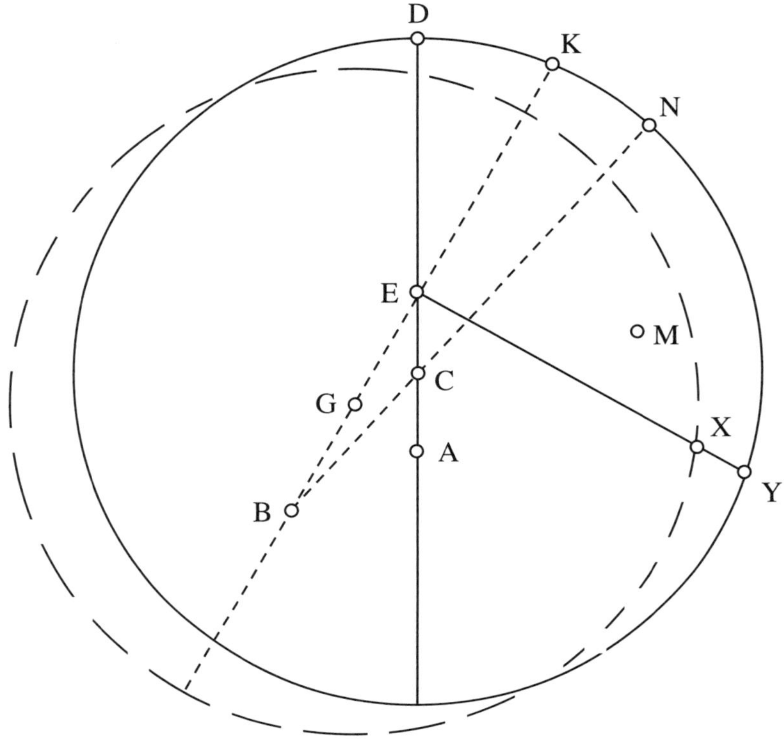

Figure 4.1. The Mean Sun versus the True Sun

until much later in the *Astronomia*. But even at this point Kepler argued that it is more physically plausible that the time variation in traversing equal arcs be proportional to distances from the sun, not some empty point (*AN*, 157–60; *KGW* III, 89–91).

The empirical inadequacy of mean sun models had not been noticed before, because mean sun models were always both constructed and tested on the basis of acronychal observations alone.[7] Only from nonacronychal observations was the discrepancy readily apparent, as Kepler spelled out in some detail in chapter 6. In figure 4.1, for example, if one were observing from M rather than from A or B, one could observe the difference between X and Y. Considering the model as it would appear from the side is an application of Kepler's requirement that hypotheses be tested against data that they were not originally constructed to explain.

This important starting point, however, was not well confirmed by empirical arguments until much later in the *Astronomia*. Indeed, though

Kepler was willing to expend a great deal of energy in constructing model after model based on the true sun, he did not consider the use of the true sun to be well confirmed empirically until chapter 52.[8] Instead, he considered his arguments to be deduced from physical considerations:

> A door is therefore open to us for deciding by means of the observations what I had deduced *a priori*[9] by consideration of moving causes: that the planet's line of apsides, which is the only line bisecting the planet's path into two semicircles equal both in size and vigor, —this line, I say, passes right through the center of the solar body and not (as the theorists have it) through some other point. In the course of the work I shall demonstrate this by means of observations, in parts IV and V. (*AN*, 167; *KGW* III, 96)

By this point, Kepler had observed that the superior planets moved more quickly when closer to the sun, which implicated the sun as the motive force (*AN*, 118; *KGW* III, 63). In addition, he considered it implausible that an empty point causally interacted with planetary bodies, forcing them to move (*AN*, 128; *KGW* III, 70). Even if the motive force were situated in the planets themselves, he did not see why they would orbit an empty point. Even if the planetary movers had minds,[10] how could they orient themselves around an empty point, and what marker could the mind use to determine its proper position in space (*AN*, 127; *KGW* III, 69–70)?[11] Kepler expanded on these physical intuitions later in the *Astronomia*, but at this point the philosophical and physical difficulties inherent in the various mean sun models were sufficient to rule them out.

Kepler also had archetypal reasons for preferring the true sun, as mentioned in chapter 2. His Trinity analogy links the sun to God, the metaphysical center of the universe, and the creative force. (Copernicus had similar metaphysical commitments to the sun, although these were not as faithfully reflected in his astronomy.) Kepler did not mention the Trinity analogy in this section, but he did comment in the introduction that he would not repeat his arguments of the *Mysterium*, but simply referred the reader to them (*AN*, 53; *KGW* III, 24). This suggests that while he still found such arguments compelling, they did not fit the subject matter of the *Astronomia*.

One wonders how much weight the Trinity analogy carried for Kepler. While the connection between the circular and the divine was a commonly accepted idea, the analogy between the Trinity and the stationary aspects of the heavens is not nearly as striking as that of the polyhedra to the moving bodies. Indeed, in the *Mysterium*, far less emphasis was placed on the Trinity than on the polyhedra. It is not obvious whether this is due to the novelty or the impressiveness of the polyhedra, although Kepler's epistemological and methodological commitments suggest the latter. The Trinity analogy has too few of the virtues that Kepler ascribed to the

Copernican and polyhedral models (fruitfulness, a striking unification of phenomena, and so on). Nonetheless, even if the Trinity analogy, by itself, carried little weight for Kepler, it does not follow that it did not play an important role in Kepler's commitment to the true sun. One thing emphasized throughout this study is the importance for Kepler of being able to draw the same conclusion from a variety of sources. Perhaps the best explanation for Kepler's confidence in the true sun, sufficient to sustain him through grueling calculations, is the convergence of support from both the physical and the metaphysical.

Later in the *Astronomia*, Kepler argued that the dignity of the sun should be considered as much evidence for its centrality as his physical and later empirical arguments:[12]

> But indeed, if this very thing which I have just demonstrated *a posteriori* (from the observations) by a rather long deduction, if, I say, I had taken this as something to be demonstrated *a priori* (from the worthiness and eminence of the sun), so that the source of the world's life (which is visible in the motion of the heavens) is the same as the source of the light which forms the adornment of the entire machine, and which is also the source of the heat by which everything grows, I think I would deserve an equal hearing. (*AN*, 379; *KGW* III, 238)

Although the sun's dignity was not given much weight in the *Astronomia*, this result appears due not to the relative insignificance of metaphysical arguments for Kepler but rather to the *Astronomia*'s particular subject matter.

Kepler's empirical arguments to the true sun have features that would please one committed to architectonic harmony. The first is that when the apsidal line of Mars is drawn through the true sun rather than the mean sun, observed planetary positions are symmetric, in both distance from the sun and angular rate. In chapter 52, Kepler followed up on the promissory note of chapter 6 and, using the distances from chapter 51, demonstrated that the aphelial point should be placed at G (in figure 4.2), that corresponding arcs from aphelion (GE and GD) on either side are symmetric in distance from the apsidal line, and that these arcs are traversed in equal times only when the apsidal line is drawn through the true sun (at point A). If, by contrast, the apsidal line is drawn through the mean sun B and point C, the arcs GE and GD, which are equal in time traversing angular distances, will be unequal in distances to the aphelial line (represented by the dotted lines BE and BD). As a result we get symmetry of distances on either side of the sun at equal arcs traversed in equal periods of time only when the true sun is used. (In figure 4.1, the solid line model is vindicated.)

While the constraint of symmetry on astronomical models was standard, one can see how this would carry extra significance for Kepler, who believed that the universe is geometrically elegant, and symmetry certainly has such elegance.

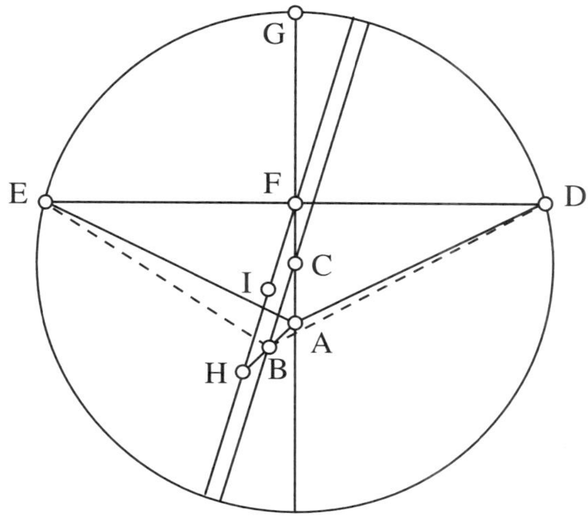

Figure 4.2. Apsidal Line through the True Sun

Kepler's second empirical argument has two parts. First, the nodes of Mars (the point at which the orbital plane of Mars passes through the orbital plane of the Earth) are in opposition to each other when taken relative to the true sun. Second, as a result of the placement of the nodes, the orbital plane is not required to librate. Kepler offered this argument in two sections of the *Astronomia* (chs. 12–14 and 61–62). His argument in the latter chapters is based on more accurate information about the planet's path: "The ratio of the orbs of Mars and the Earth, the eccentricity of each, and the shape of their paths, have all been found with great certainty in the preceding chapters. Therefore, we can now easily accomplish here what we sought out in an approximate way in chapters 11, 12, 13, and 14" (*AN*, 605; *KGW* III, 385). In chapter 61 he determined that in 1595 the nodes were at 16°46⅓′ Taurus and 16°46⅓′ Scorpio, exactly 180° apart (*AN*, 606; *KGW* III, 386). When taken this way, the nodal line passes through the body of the true sun. If the model is constructed on the basis of the mean sun, then the nodes are not in opposition. Instead they are at 17°38′ Taurus and 15°44½′ Scorpio (*AN*, 631; *KGW* III, 401–2). Kepler thought this failure of opposition an implausible state of affairs: "For why will the diameter of the intersection of the planes not intersect the diameter of the apsides in the center from which the eccentricity originates . . . ?" (*AN*, 631; *KGW* III, 402).

These arguments, that the inclination of the plane of the orbit is stable, and is approximately 1°50′, Kepler followed up in chapter 62. His fixing of the nodal line in chapter 61 did not require that the plane librate, an additional argument that the plane indeed does not librate, and further support for the accuracy of his placement of the nodes. In chapter 14 he used three different measures of the inclination, where Mars is close to conjunction with the sun, where it is in quadrature with the sun, and where it is in opposition to the sun. In chapter 62 Kepler repeated his argument with observations of Mars in opposition to the sun. Apparently he felt it unnecessary to supply arguments based on conjunction and quadrature, simply encouraging readers to carry out the argument if they remained dissatisfied (*AN*, 611; *KGW* III, 389). Presumably he felt confident that by this point his arguments had been sufficiently persuasive.

The symmetry (of times taken to traverse equal arcs and of distances) and elegance (stable inclination) of the true sun models would appeal to one who, like Kepler, believed that the universe was constructed to be aesthetically pleasing. Kepler at this point also had two *independent* empirical arguments for the use of the true sun rather than the mean. The position of the nodes is independent of the position of the apsidal line. One does not determine the other. Two independent and aesthetically pleasing empirical arguments for the true sun, in addition to Kepler's archetypal and physical arguments, were persuasive support indeed.

The Vicarious Hypothesis

Kepler constructed the first model based on the true sun in chapter 16. He still believed orbits to be circular, so to determine the orbit he needed to specify the placement of the apsidal line, the eccentricities of sun and equant, the size of the orbit, and the mean anomaly of one observation (the proportion of the period and time taken to traverse from aphelion to that observational position). These parameters, specified using four acronychal observations (oppositions), determine the angles from the sun at A in figure 4.3, as well as the angles from equant point C, since motion around the equant is uniform.

To determine the eccentricity of the sun to the center (AB) and the eccentricity of the equant (BC), Kepler assigned values for angles FCH (the mean anomaly) and FAH (the true or coequated anomaly). He tested each assignment of values by trigonometrically determining whether the points F, G, D, and E lie on a circle, and whether B lies on a line between A and C. If not, then he assigned different values and started again. This was no easy task: "If this wearisome method has filled you with loathing, it should more properly fill you with compassion for me, as I have gone through it at least seventy times at the expense of a great deal of time, and you will cease to wonder that the fifth year has now gone by since I took up Mars" (*AN*, 256; *KGW* III, 156).[13] The result was a model in which if the radius is taken to be 100,000 units, then the whole eccentricity (AC) is

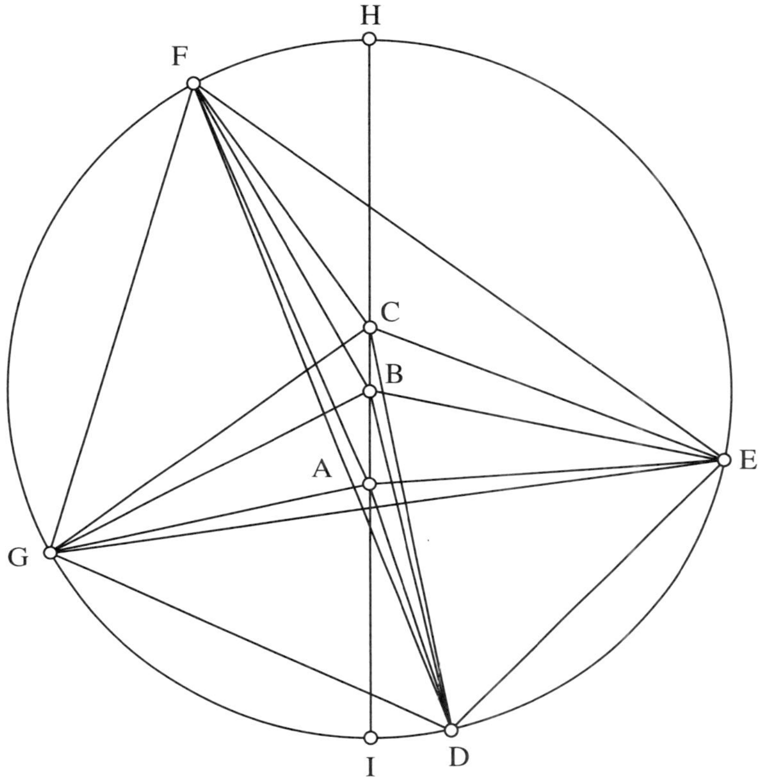

Figure 4.3. The Vicarious Hypothesis

18,564, the eccentricity of the sun to the center (AB) is 11,332, and the equant to the center (CB) is 7,232 (*AN*, 269; *KGW* III, 269).

In chapter 18 Kepler tested the vicarious hypothesis against twelve acronychal observations, finding it accurate within about two minutes.[14] Nonetheless, in chapters 19 and 20 he immediately refuted the vicarious hypothesis against observations of a different type from those on which he had originally constructed the model. Kepler used two independent means to show that the eccentricity in the vicarious hypothesis was incorrect. In chapter 19 he used acronychal latitudes to determine that the distance from the sun to the center (AB) should be between 8,000 and 9,943 (nearly a bisected eccentricity) rather than the 11,332 determined by the vicarious hypothesis (*AN*, 281–83; *KGW* III, 174–76). A bisected eccentricity gave better distances, but was not as good for the twelve longitudinal positions used for testing the vicarious hypothesis in chapter 18. The error in heliocentric longitude introduced by a bisected eccentricity was about eight minutes for positions near the octants, which Kepler found unacceptable.

Although an eight-minute error had been acceptable to previous astronomers, Kepler was sure that Brahe's data were accurate to within two minutes of arc. As many of Kepler's commentators have noted, this refusal to accept an eight-minute error was a pivotal moment. Had Kepler accepted it, ending the *Astronomia* with the vicarious hypothesis, this would have been a significant achievement, for the hypothesis was more accurate than previous models. But it pales in comparison to the contributions Kepler eventually made, a reform of astronomy motivated by the erstwhile acceptable eight-minute error. Kepler credited this turning point, in large measure, to Brahe,

> since the divine benevolence has vouchsafed us Tycho Brahe, a most diligent observer, from whose observations the 8' error in this Ptolemaic computation is shown. . . . Now, because they could not have been ignored, these eight minutes alone will have led the way to the reformation of all of astronomy, and have constituted the material for a great part of the present work. (*AN*, 286; *KGW* III, 178)

Although this statement suggests that Kepler had already rejected the vicarious hypothesis, he immediately offered another argument against it. In chapter 20 Kepler used nonacronychal observations, where the Earth is orthogonal to the sun and Mars, to determine the distance from sun to center (AB) to be somewhere between 8,377 and 10,106, rather than 11,332 (*AN*, 287–90; *KGW* III, 178–80).

In light of the discussion of the *Mysterium* and the *Apologia*, it is clear why Kepler rejected the vicarious hypothesis. It was accurate enough for the phenomena on which it was based but not for related phenomena.[15] Although one could save the related phenomena by constructing separate models for longitudes and latitudes, such models could not be consistently unified in a single orbit. The vicarious hypothesis lacked the fruitfulness and unity that Kepler praised in the Copernican model. As noted earlier, because these archetypally grounded theoretical virtues were necessary to distinguish between mere empirical adequacy and truth, the vicarious hypothesis had to be rejected as a model of the actual Martian orbit.

Kozhamthadam (among others) observes that:

> Perhaps in no other instance is his [Kepler's] emphasis on observation or empirical inquiry more conspicuous than in the discovery [and rejection] of the vicarious hypothesis. . . . Kepler's emphasis on observations to arrive at the correct astronomical theory was even more striking in his refusal to accept the time-honoured bisection of the eccentricity.[16] . . . What would have been more har-

> monious and symmetrical than having the eccentricity equally divided by the center of the orbit? (Kozhamthadam1994, 166–67).

Kozhamthadam nonetheless refuses (rightly) to conclude that Kepler's cosmological views became less important or unimportant. Indeed, they became quite important in the physical exploration and justification of the ellipse.

The New Physics

In the introduction to the *Astronomia* Kepler implied that the connection between the geometrical and physical levels of the universe justifies, and indeed requires, the conjunction of astronomy and physics:[17]

> Indeed, all things are so interconnected, involved, and intertwined with one another that after trying many different approaches to the reform of astronomical calculations, some well trodden by the ancients and others constructed in emulation of them and by their example, none other could succeed than the one founded upon the motions' physical causes themselves, which I establish in this work. (*AN*, 48; *KGW* III, 20)

Before Kepler could effectively use physics to ground astronomy, he had to discover the criteria for a plausible physical explanation. Traditional Aristotelian physics was not suitable. Kepler argued that geometrical points cannot be physically efficacious, and, as a result, the center of the universe, qua geometrical center, cannot account for the movements of physical bodies. Only centers qua centers of physical bodies can attract other bodies (*AN*, 54–55; *KGW* III, 24–25).

One such criterion is a consequence of Kepler's conception of the relationship between God, the human mind, and the physical universe. In the *Mysterium* Kepler claimed that an implication of our creation in God's image is that our minds can recognize the archetypes. In April 9/10, 1599, Kepler reexpressed this belief in a letter to Herwart von Hohenburg:

> To God there are, in the whole material world, material laws, figures and relations of special excellency and of the most appropriate order. . . . Let us therefore not try to discover more of the heavenly and immaterial world than God has revealed to us. Those laws are within the grasp of the human mind; God wanted us to recognize them by creating us after his own image so that we could share in his own thoughts.[18] (Baumgardt 1951, 50; *KGW* XIII, 308–9).

Kepler used this to delimit the kinds of physical hypotheses that could count as plausible. Given the mirroring between archetypal and physical structures, physical hypotheses that "disturb the mind"[19] can be ruled out.

By "disturbing the mind" Kepler meant something like this. If we cannot think of an event occurring unless we think of it as occurring in time, then the concept of events and the concept of time are linked together. If we cannot think of motion without thinking of it occurring in space, then the concept of motion and the concept of space are linked together. As a result, to try to conceptualize events without time, or motion without space, is to think in a way that disturbs the mind. This thesis is more than just psychological for Kepler; it is ontological as well, because of the connection between our minds and the archetypes, and between the archetypes and the physical world.

The criterion of comprehensibility played a significant role in Kepler's argument that the sun causes the planets' variation in speed, for it allowed Kepler to narrow the range of possible causes to either the planets or the sun. Kepler noted that speed is correlated with distance and suggested that either one is the cause of the other, or some third cause is the cause of both. But Kepler ruled out a third cause on the grounds that it would be impossible to conceive of such a cause (*AN*, 376; *KGW* III, 236).

One wonders why Kepler said this. One could certainly come up with a more convoluted story. For example, perhaps for archetypal reasons the density of a planet increases the further it is from the sun, and the difference in density is the cause of the variation in velocity but not the cause of the distances. In this example, the archetypes cause both the distances and, indirectly, the velocities. The direct cause of velocity is unrelated to distance. Kepler later considered the possibility that planets vary in density, resulting in varied resistance to the sun's force (*Epitome*, 880; *KGW* VII, 283). Be that as it may, accepting the impossibility of imagining some third cause allowed Kepler to move with confidence to focusing on the remaining two possibilities. Here it is plausible to say that Kepler had the goal in sight, and his focus was on justifying rather than exploring the main cause of planetary motion.

In chapter 33 Kepler argued that the variation of distance must be the cause of the variation in motion, rather than the reverse, because distance is prior conceptually to motion.

> For distance from the center is prior both in thought and in nature to motion over an interval. Indeed, motion over an interval is never independent of distance from the center, since it requires a space in which to be performed, while distance from the center can be conceived without motion. Therefore, distance will be the cause of intensity of motion, and a greater or lesser distance will result in a greater or lesser amount of time. (*AN*, 377; *KGW* III, 236)

In other words, to conceptualize motion without thinking of distance is to think in a way that disturbs the mind. Distance, however, can be conceptualized without reference to motion and is therefore prior to motion. As a distance is defined by its end points, the cause of planetary motion must be at one of the end points. That is, either the planet or the sun is the cause of the planet's motion (*AN*, 377; *KGW* III, 236).[20]

In chapter 33 Kepler presented various arguments that the primary motive force is provided by the sun, not the planets. He argued that the planets' speed varies relative to their distance from the sun, that the apsides and orbital plane of Mars are best represented as drawn through the sun, and that the sun is the most dignified heavenly body (*AN*, 378–79; *KGW* III, 237–38).

Once again, Kepler used his archetypes to justify known results. He already believed that the sun was the primary motive force, and his purpose here was to persuade the reader that the philosophical foundations of such a claim were solid. He needed to offer an alternative to the Aristotelian conception of the essential characteristics of heavenly bodies in motion. That is why he emphasized that his physical approach was both comprehensible and, given the sun's dignity, appropriate.

At this point, however, Kepler had not yet offered an explanation of the efficient causes of planetary motion. But he thought that any study of the physical universe must at least point toward efficient causes, and that the nature of these is material. There are two aspects of Kepler's materialism. First, he aimed at producing a physical theory of the universe that would describe the universe as a mechanical system:

> My goal is to show that the heavenly machine is not a kind of divine living being but similar to a clockwork in so far as almost all the manifold motions are taken care of by one single absolutely simple magnetic bodily force, as in a clockwork all motion is taken care of by a simple weight. And indeed I also show how this physical representation can be presented by calculation and geometrically. (Letter from Kepler to von Hohenburg, February 10, 1605; quoted in Caspar [1959] 1993, 136; *KGW* XV, 146)

Second, he was guided by the idea that celestial physical causes are similar to terrestrial physical causes. This similarity allowed him to speculate on the nature of celestial causes by drawing on analogies with more ordinary Earthly causes. The method of uncovering causal mechanisms by analogy with known forces is evident, perhaps paradoxically, in Kepler's discussion of the possibility of planetary minds as guiding forces.

Kepler argued that even with minds, planets need some external indication of where they are in orbit and how fast they are going because they have to speed up and slow down accordingly. Yet the only indicator available to the planet is the sun's apparent diameter, as the eccentric circle and epicycle alike are centered on empty points. Just as sailors cannot orient themselves on the ocean without physical indicators of where they are, neither can planetary minds in orbit (*AN*, 412–13; *KGW* III, 260). Even if one supposed that a planetary mind could observe the apparent changes in the sun's diameter, and use that as a marker, it would also need to bear in mind the measure of libration, and these changes cannot be unified by a simple proportion:

> It should be known, however, that the increases of the diameter of the sun and the arcs of the epicycle do not square with each other well, and so the motive mind will have to have a very good memory. (*AN*, 415; *KGW* III, 262)

> You see, my thoughtful and intelligent reader, that the opinion of a perfect eccentric circle for the path of a planet drags many incredible things into physical theories. *This is not, indeed, because it makes the solar diameter an indicator for the planetary mind, for this opinion will perhaps turn out to be closest to the truth,*[21] *but because it ascribes incredible faculties to the mover*. (*AN*, 416; *KGW* III 262; emphasis added).

The mental power required of the planet is incredible, for it must remember which libratory movements correspond to which apparent solar diameters. This task is difficult because the two do not covary if the orbit is assumed to be circular. For Kepler, if the appeal to planetary minds is to explain efficient causation, then one cannot ascribe supernatural powers to them. If there must be minds, then we need to explain how they are able to perform their function naturally, and so physical explanations will involve reference to some body regardless.

Although these arguments favor his hypothesis over the Ptolemaic (recall how epicycles require the planet to orient itself around an empty point), Kepler preferred analogies based on physical causation. While working on the *Astronomia*, he also wrote the *Astronomiae pars optica* (1604).[22] Not only did this study provide Kepler with an account of refraction and its effect on astronomical observations, it also provided an example of how a body can exude a force acting on distant bodies.

Given the influence of Neoplatonists like Plotinus on Kepler, it was natural for him to draw analogies between light emanating from the sun, motive force emanating from the sun, and creative force emanating from God.[23] The analogy between light and motive force was especially apt—not only did they relate to the Trinity, Kepler believed, but they related to physical objects in much the same way.

Kepler proposed that motive force, although not corporeal, has the ontological status of being potentially physical: "exactly like light, between the source and the movable thing it is in a state of becoming, rather than of being" (*AN*, 382; *KGW* III, 240). Kepler argued that "the *species* can exist or subsist nowhere in the world but in the bodies of the mobile things themselves" (*AN*, 382; *KGW* III, 240), that the motive force takes no time at all to travel distances, and that it loses none of its power over distances. Nonetheless, he claimed that since the motive power originates from physical bodies and acts on them, it is subject to geometrical laws just as light is. As a result, the effect of motive power attenuates with distance, as it spreads out over a larger area. The spreading, rather than the loss of power, explains the inverse proportion of the planet's speed to its distance from the sun (*AN*, 381–83; *KGW* III, 240–41).[24]

Kepler noted that although light served as a useful analogy, magnetic force was even better: "The example of the magnet I have hit upon is a very pretty one, and entirely suited to the subject; indeed, it is little short of being the very truth" (*AN*, 390; *KGW* III, 246). Kepler felt that an exploration of magnetic forces was promising, already believing William Gilbert's claim that the Earth is a giant magnet: "It is therefore plausible, since the Earth moves the moon through its *species* and is a magnetic body, while the sun moves the planets similarly through an emitted *species*, that the sun is likewise a magnetic body" (*AN*, 391; *KGW* III, 246).

Magnets have several parts, including an attracting force, a repulsing force, and a directing force that orients metal strips in particular patterns. Kepler thought it plausible that the motive force of the sun is not an attracting or repulsing force, but a kind of directing force, a transverse pushing force (*AN*, 390; *KGW* III, 246), while the libration is explained by attraction and repulsion between the sun and the planet's magnetic poles.

Kepler proposed that since the planets move around the zodiac, the magnetic fibers are set perpendicular to the poles of the sun. The sun spins on its axis, flinging rays of magnetic power in straight lines out in a circle. These rays grab on to the planets and push them in a circle close to the ecliptic. The planets only move in one direction because the sun only spins in one. The planets stay near the ecliptic because the filaments at the poles are not suitable for moving the planets around. Figure 4.4 illustrates this nicely. In the left-hand figure, the sun is as it appears to a planet on the ecliptic. All the magnetic fibers are pointing in a single direction, pushing the planet in one direction. The right-hand figure shows how the sun would appear if the planet was high above the ecliptic. As the planet moves

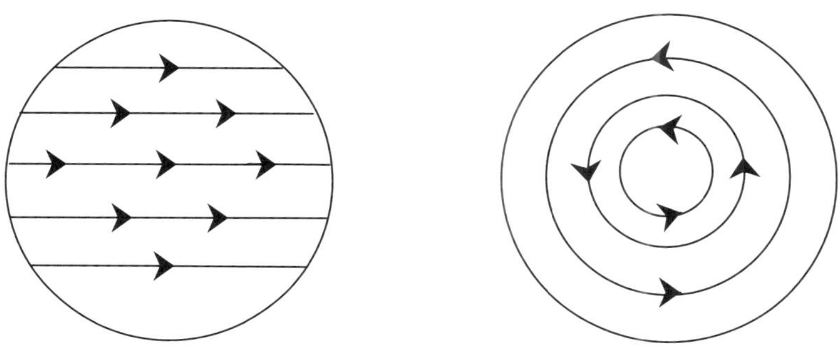

Figure 4.4. The Magnetic Rays of the Sun

farther away from the ecliptic, the moving force becomes more complicated, with fibers pointing in opposite directions. This arrangement, Kepler stated, is unsuitable for moving the planets around (*AN*, 398–99; *KGW* III, 251–52).

The magnetic analogy, as developed thus far, only explains how the planets move in a circle. It does not explain their movement toward and away from the sun. Kepler proposed that the planets must exert some kind of force and suggested that the power that the planets exert produces movements similar to those of a boat when an oar is used to steer the boat (*AN*, 404–5; *KGW* III, 255). Kepler used this analogy to explain how something can interact with a force to alter its effect, and to explore what properties a planet's powers would have to have to interact with the motive power from the sun and produce the appropriate motion.

All of these analogies share a common feature. They are based on the assumption that celestial activity is similar to the terrestrial (in the case of planetary minds, the supernatural is naturalized). Although the drive for explanatory unification was not new with Kepler (recall Thales), there was a strong tradition to view the celestial and the terrestrial as distinct, just as the divine and the physical were not to be conflated. While Kepler adhered to the latter distinction, he did not think the divine began at the borders of the Earthly. It belonged rather to an entirely different plane (the archetypal). Anything accessible to the senses is not divine but a material representation of it. As a result, the physical universe includes celestial bodies, and these bodies are subject to the same principles as other physical bodies, because they were all created to instantiate the archetypes. For Kepler, then, his archetypal universe permitted the exploitation of knowledge of terrestrial causes without requiring the more radical abandonment of the distinction between the divine and mundane.

By the end of chapter 39, Kepler had a sketch of what he expected the new physics to look like and, perhaps more important, an account of what it could not be. The new celestial physics could not be mysterious, it could not invoke empty points in space in causal explanations, and it could and perhaps should be commensurate with terrestrial physics.

Kepler's Second Law

Before we discuss the second law,[25] a review of terminology is in order. Kepler spoke of "delay" rather than "velocity," where the delay of a planet is the time it takes to span a small arc of its orbit. The slower the velocity, the larger the delay. Stephenson argues that Kepler used delays rather than velocities because delays could be represented geometrically, whereas velocities could not. Here Stephenson (1987, 62) refers the reader to Euclid, noting that velocities cannot be represented as proper ratios that could be represented geometrically. While this seems right, Stephenson does not comment on the likely significance of the archetypes here. Given that Kepler saw the universe as ordered by geometrical archetypes, a nongeometrical representation of a physical aspect of an orbit would not be amenable to archetypal substantiation. In the next section I discuss how Kepler made the archetypal significance of geometrical representation explicit.

In chapter 32 Kepler argued that the physical interpretation of the equant model implies that for equal arcs at the apsides, the delays are proportional to the distances from the sun. This proposal is known as Kepler's distance law, though he did not refer to it as such. Kepler closed chapter 32 by noting that the distance law did not exactly match the equant model when outside the apsides.

At the end of part III, Kepler addressed the issue of how to determine the position of the planet in its orbit. The distance law did not provide an easy answer, for the time taken to traverse a given arc varied with distance from the sun, and so determining the position required a summation of an infinite number of individual distances. To cope with this problem Kepler devised a geometrical calculation substitute. He divided the circle into one degree segments around its center B (figure 4.5), forming sectors CGB, GHB, and so on. Because the altitudes and areas of the triangles within each sector are equal, and the arcs are equal, the areas of the sectors can be taken for the arcs, and can be used as a measure of the distances they contain, and of the times taken to traverse the arcs. Furthermore, because the distances from the sun at A to the circumference are in the same circle as those from B, the sum of the areas measures the sum of the distances to the sun that the areas contain. This insight led to what today is called Kepler's second law (the area law), that the line between the planet and the sun (the radius vector) sweeps out equal areas in equal amounts of time.

Kepler, however, realized that the areas from the center of a circular orbit did not precisely measure the distances from the sun in a circular orbit, because the sectors constructed with equal arcs CG, GH, and so on, have unequal bases if the sun rather than the center of the circle forms the third vertex. To demonstrate the discrepancy, Kepler devised a figure that was constructed by unfolding half of the circular orbit (figure 4.5) and representing the radii as parallel (CB, GB, and so on). In figure 4.6a, the line CGHEIKD is the unfolded circumference of the half-circle CED. The line BBBBBBB simply represents the center of the orbit B, with distance CB in figure 4.5 corresponding to CB in figure 4.6a, GB to GB, and so on. Area CGBB in figure 4.6a is proportional to CGB in figure 4.5, though the rectangle is twice the size of the triangle, since B was "stretched out" to form two corners of the rectangle. As a result, the area of the figure in figure 4.6a is proportional to the areas from the center B in figure 4.5.

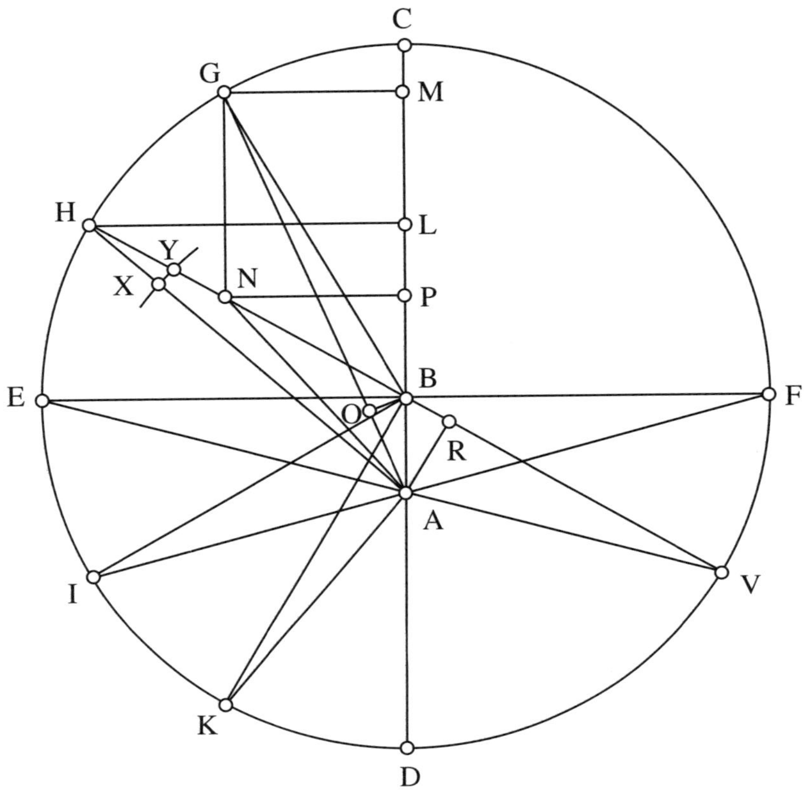

Figure 4.5. The Circular Orbit for the Areas

The areas from the sun A are represented by the conchoid in figure 4.6b, where A is stretched out over the curve AAAAAAA. Area CGAA in figure 4.6b is proportional to area CGA in figure 4.5.

But notice that the areas of triangles from the sun A represented in figure 4.6b are not equal to those from the center B represented in figure 4.6a. The figure that *would* be equal is the one in figure 4.6c. This measures the sum of the distances CA, HR, and so on, which are constructed in figure 4.5 by taking the perpendicular from A to the line from the point on the circumference through B. Kepler showed us this construction for H, where the perpendicular from A to the line through H and B is at R in figure 4.5. The same can be done for Q, S, and L in figure 4.6.

The discrepancy between the two figures (figures 4.6b and 4.6c) gave Kepler an idea of how the sums of areas from the center (figure 4.6a) differed from the sums of distances from the sun (figure 4.6b). Interestingly

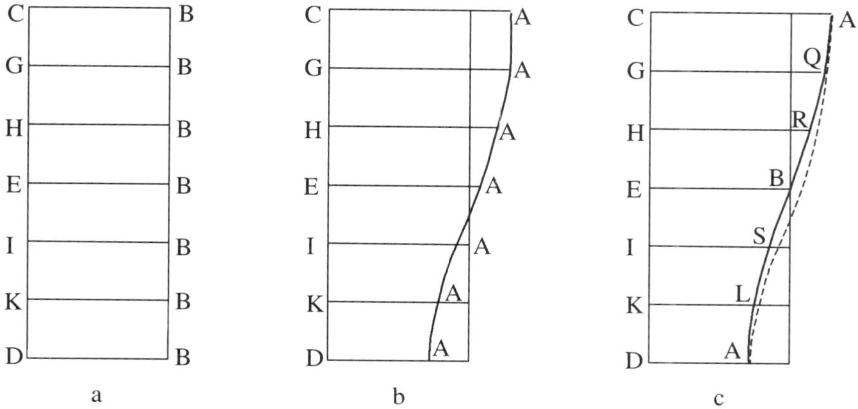

Figure 4.6. Kepler's Conchoids

enough, it turns out that the conchoid in figure 4.6c measures the distances from the sun in an elliptical orbit. In a circular orbit, however, this conchoid is physically peculiar, as it measures distances to a point off the sun (R, Q, and so on), rather than distances to the sun. As characterized in chapter 40, the orbit based on this conchoid is physically implausible.

Despite this, the means of measuring sums of distances by the sums of areas has something going for it that the distance law does not; the areas described are symmetric to the center of the conchoid (figure 4.6c), whereas the conchoidal depiction of the distance law is not. There is now a conflict between Kepler's different notions of plausibility. The area rule has features that make it archetypally plausible, whereas the distance law was physically motivated. If Kepler still believed at this point that the geometrical and the physical mapped onto each other, then the discrepancy between the area rule and distance law should have been troubling. Kepler gave little evidence of being troubled at this point. So if anything, Kepler's area rule is a counterexample to my thesis that he still required astronomy to conform to the archetypes. One should bear in mind, however, that Kepler had not yet completed his war on Mars, and his triumph would eventually involve a geometrically elegant account of its orbit.

Oval Orbits

One wonders how difficult it must have been for Kepler to give up the circular orbit. Not only was he going against a time-honored tradition (which admittedly he was willing to do in other cases); he was also giving up the calculational ease of circles, and abandoning an obvious archetypal explanation for the shape of orbits. A study of the *Harmonice* reveals he was unwilling to suppose that the physical orbit simply deviated from its archetypal shape.[26] One theme of this study is that Kepler was committed to the idea that the physical replicates perfectly the divine, without deviation. I am not aware whether he had at this time conceived of harmonic ratios as filling the gap. He did, however, conceive of harmonic ratios as early as 1599 as potentially explaining the periodic times of the planets (Kepler wrote of this in a letter to von Hohenburg, August 6, 1599; *KGW* XIII, 21–41). But even if he had, he did not have sufficient reason to be confident that an archetypal explanation for elliptical nonuniform motion was forthcoming. Kepler had not yet obtained the wonderful convergence of results that would have given him such confidence. Nonetheless, he was willing to struggle with noncircular orbits, a task so frustrating that at times he "was straining and gnashing . . . [his] teeth" (*AN*, 458; *KGW* III, 290). At this point, the noncircular orbit was determined a posteriori (*AN*, 455; *KGW* III, 288), which for Kepler amounted to a "lucky guess."[27]

To determine the placement of the orbit, Kepler needed to locate the circle that would pass through the apsidal line (and be symmetrical to it) as well as three observed positions. In chapter 41 he found that once the apsidal line was placed, no circle was consistent with all three observed positions. In chapter 42, Kepler determined the apsidal line more precisely, and used this line in chapter 43 with a bisected eccentricity and the area rule to determine where the planet would be at a given time. By triangulations he determined that his theory made the planet move too slowly near the apsides and too quickly in the mean distances. He argued that the area rule was not at fault because, given its discrepancy with the distance law, it would have erred in the opposite direction (*AN*, chs. 41–44). Because the area rule was not at fault, the orbit had to be noncircular.

Once he determined that the orbit is not circular, he attempted to determine a model for the noncircular orbit, which he believed to be an oval (Kepler mentioned the need for an oval orbit in a letter to Fabricius on October 1, 1602; *KGW* XIV, 278). Here Kepler's metaphysical commitments are at the fore, quite possibly because the oval is of lesser divinity than the circle, and so its selection required metaphysical justification. This concern is suggested by Kepler's rejection of the first physically defined ovoid model of chapter 45. The model was constructed from two principles: that the epicycle turns uniformly, and that the motion of the planet around the sun is nonuniform but proportional to the distances. The problem was that Kepler could not generate the orbit geometrically. One cannot determine the length of the arc traversed at a given time unless one knows the total length of the orbit, which in turn cannot be determined unless one knows the length of the arc traversed at a given time.

> This, as you see, begs the question, and in our operation we presupposed what was being sought, namely, the length of the oval. This is not just a fault in our understanding, but is utterly alien to the primeval ordainer of the planetary courses: we have hitherto found no anticipation of such lack of geometry in the rest of his works. (*AN*, 489; *KGW* III, 310)

Because it did not have the right geometrical match, the ovoid model could not comport with an archetypal account. Stephenson (1987, 103) notes that there is nothing physically problematic with the first ovoid model, and that its rejection displays a greater interest in archetypal models than in physical models . On my interpretation it is *precisely* a concern with physical understanding that motivated the rejection. Kepler believed that it is the possibility of correspondence to an archetypal model in virtue of which one could be confident that a physical model is correct. My contention is that viewing his archetypal approach as an alternative to physics obscures the purpose of developing aesthetically pleasing geometrical models, to wit, constructing the true physical astronomy.

Because of the difficulties with the first oval, in chapter 49 Kepler offered a modified epicycle model where the center of the epicycle, rather than the planet, moves nonuniformly, at a speed inversely proportional to distance from the sun (*AN*, 490–92; *KGW* III, 311–12). This model was less empirically adequate than the first epicycle model, and it was also inconsistent with Kepler's denial of the physical reality of epicycles, as he commented in a side note (*AN*, 491; *KGW* III, 311–12). In the first ovoid model the epicycle can be used as a mathematical device, whereas in the second it cannot because there its center plays an explanatory role. In the modified oval, the motion of the center of the epicycle, rather than the planet, is determined by distance from the sun, which means that the sun's motive force would have to act on an empty point in space rather than on a body. Thus, although the second model was geometrically definable, it did not fit Kepler's requirements for physical plausibility and empirical accuracy.

Exploring these models did not prove fruitless, however. In doing so, Kepler determined the proportion of the lunula (a figure in the shape of a narrow crescent, like a crescent moon) that had to be sliced from the circle. This calculation led to a rather serendipitous discovery. He needed the lunula to be about 429 parts at the widest section and noticed that in an orbit with a radius of 100,000, the secant of the angle measuring the greatest optical equation is 100,429. The optical equation is the angle sun-planet-center, angles AGB, AHB, AEB, and so on, in figure 4.7. The optical equation is greatest around E, where the widest slice of the lunula needed to be. Kepler suggested that the radius be substituted for the secant to produce the correct orbit. What he meant by this is that distances from

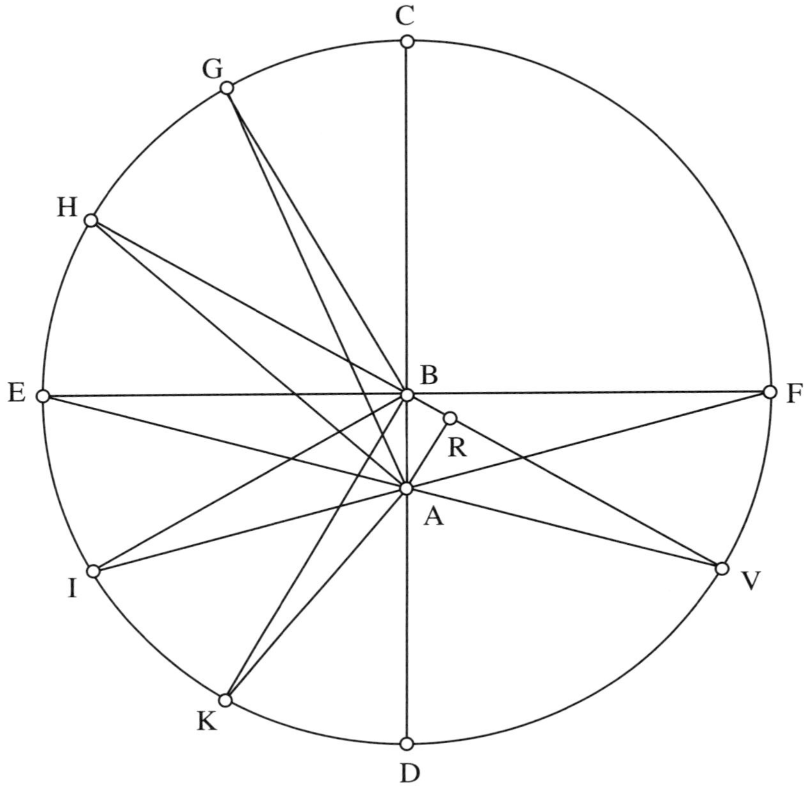

Figure 4.7. Determining the Proportion of the Lunula

the sun GA, HA, EA, and so on should be the lengths of lines constructed by dropping a perpendicular from A to the line through the planet and the center of the orbit. The distances from the sun are thus determined by HR rather than HA, EB rather than EA, and so on in the same manner that the distances for the conchoid in figure 4.6c were determined.

Again, for the same reason that Kepler's conchoid in figure 4.6c is peculiar, these distances are physically peculiar because they are not determined as distances between bodies. That is, while the distances constructed in this manner are distances between the planet and the sun, they are not *constructed* as distances between the planet and the sun. The method of construction is not physically motivated and does not naturally suggest a physical explanation for the orbit. At this point Kepler had not finished investigating the causes of Mars's motion. He needed an accurate physical account of what caused the planet to librate.

Physical and Geometrical Models of Libration

Kepler observed at the opening of chapter 57 that once he had determined the orbit to be noncircular, his physical analysis would proceed more accurately. He stated that the libration of the planet varies with the versed sine of the eccentric anomaly (angle DBC in figure 4.8). Kepler spoke as though he was offering a physical account of libration: "The reciprocation [libration] is coordinated with the space traversed on the eccentric by natural means, which depend not upon the equality of the angles DBC, EBD, FBE, but upon the strength of the ever increasing angle DBC, EBC, FBC" (*AN*, 547; *KGW* III, 348). He added in a side note that the strength of the angle at the center of the orbit (B) "is the reason why the versed sine of the eccentric anomaly is the measure of this reciprocation [libration]" (*AN*, 547–48; *KGW* III, 348).

Although Kepler spoke in this manner, he was well aware that the versed sine measure was a geometrical rather than a physical account. The eccentric anomaly is the angular distance of the planet from aphelion, measured by the angle at the center of the orbit. If the planet is at D, the eccentric anomaly is measured by CBD. This center is a geometrical point, rather than a physical body, and so, in accordance with Kepler's physical principles, the versed sine of the eccentric anomaly cannot be a physical measure. Again, for Kepler geometrical points could play no role in physical explanations, as they are not causally efficacious.

Kepler was not confused, he was merely confident in the versed sine measure. It is the first elegant geometrical formula he had been able to derive after rejecting the circular orbit. Holton (1956, 346) interprets this lack of distress as an indication of Kepler's lack of commitment to physical explanations, because he had an alternative geometrical model to fall back on. My claim is that given Kepler's belief in the mirroring of domains, an elegant geometrical measure suggests a corresponding physical cause, and so he felt confident that an adequate physical model was forthcoming: "And since the finger points to a natural way of measuring this reciprocation [libration], its cause will also be natural; that is, it will be some natural—or better, corporeal—faculty, and not a planetary mind" (*AN*, 548; *KGW* III, 348). In other words, if a simple equation can describe such complex motion, then it is likely a simple physical mechanism that caused it. This claim implies that Kepler continued to believe in the mutual mapping of geometrical and physical models, and that geometrical regularities indicate physical causes. This is very different from seeing geometrical models and metaphysical models as mere backup positions.

Kepler's next step was to explore a physical account of libration. He thought that conceiving the planets as magnets held the key to uncovering

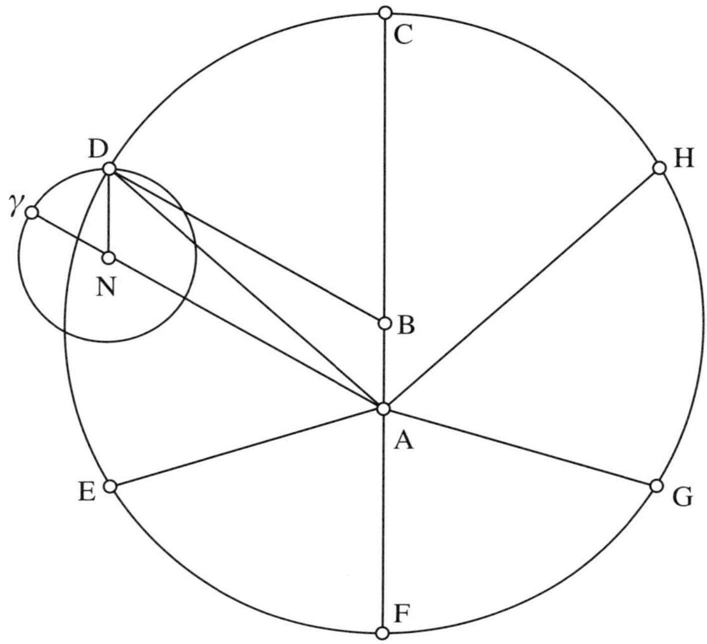

Figure 4.8. The Strength of Libration

the causes of libration: "But come: let us follow more closely the tracks of this similarity of the planetary reciprocation [libration] to the motion of a magnet, and that by a most beautiful geometrical demonstration, so that it might appear that a magnet has such a motion as that which we perceive in the planet" (*AN*, 554; *KGW* III, 352). In Kepler's "beautiful demonstration," the magnetic axis of the planet is represented by DA, and the sun is on the line through BK, in the top figure of figure 4.9. D is the pole attracted to the sun, and A is the pole repulsed. Angle CBI is the coequated anomaly (and, hence, so is PCB) measured by arc IC. The coequated anomaly, or the true anomaly, is the angular distance of the planet from aphelion as measured by the angle at the sun, angle CAD in figure 4.8. Arc IG measures the eccentric anomaly, which is CBD in figure 4.8. Angles DBK and ABK indicate the exposure of attractive and repulsive poles to the sun, and the relative sizes or "strengths" of these angles measure the attractive and repulsive forces. Kepler's next move was to assume that the effects of the strength of angle DBK, against the strength of angle ABK, would work according to the law of balances. Kepler did not explain this assumption, saying only that "since the strength of this angle is natural, it will follow the same ratio as the balance" (*AN*, 556; *KGW* III, 353). Presumably he was

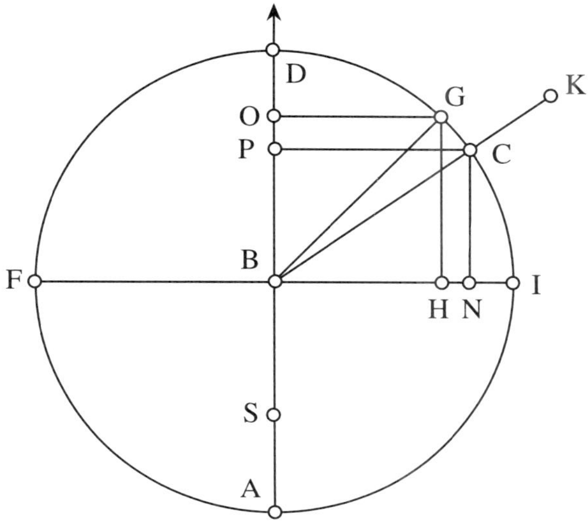

Figure 4.9. Kepler's Beautiful Magnetic Demonstration

following the tradition of mechanics in assuming that mechanical movements were reducible to the balance.[28] If the system is in equilibrium in the position shown,[29] then it can be reduced to a balance in equilibrium with arms DP and AP suspended from CP (this is clearest if figure 4.9 is rotated 45 degrees so that CP is vertical). This result occurs because the ratio of strengths of ∠DBK:∠ABK is proportional to the ratio of weights D:A, and therefore inversely proportional to lengths DP:AP. Because DP measures the repulsive force, and AP the attractive force, the net force will be the difference between DP and AP, which is SP, constructed by marking AS as equal to DP. SP is always twice PB, which is equal to CN, the sine of the coequated anomaly. Therefore, at a given time the strength of the planet's attraction to, or repulsion from, the sun is measured by the sine of the coequated anomaly (*AN*, 555–56; *KGW* III, 352–53).

The sine of the coequated anomaly is a candidate for counting as a physical measure, because the coequated anomaly is the angular distance of the planet from aphelion as measured at the sun. That is, the vertex of the angle of the coequated anomaly is at the center of a body, and not simply an abstract mathematical point. Although this sine may be a measure of a physical force, it did not by itself constitute a physical theory for Kepler, because he believed that such theories are qualitative descriptions of the causal mechanisms involved.

So far Kepler had explained only what the strength of the magnetic force would be at a given time. His next move was to account for the cumulative effect over time, and to argue that this would be equivalent to the versed sine of the eccentric anomaly:

> The measure of the distance of the reciprocation [libration] traversed by these continuous increments of power is quite another thing. For the observations show that if the eccentric anomaly GI corresponds to the equated anomaly IC, the versed sine IH of the arc GI is the measure of the reciprocation [libration] accomplished [figure 4.9]. If this can also be deduced from the previously indicated measure of the speed CN, then we shall have reconciled experience with the demonstration involving the balance. *Since the sine of any arc is the measure of the strength of that angle, the sum of the sines will be an approximate measure of the sum of the strengths or impressions over all the equal parts of the circle.* (*AN*, 556–57, *KGW* III, 353–54; emphasis in original)

He noted that while the two measures were not precisely equivalent, the errors were negligible (*AN*, 559; *KGW* III, 355). What Kepler showed is that the sum of sines up to a given point on the circle is equivalent to the versed sine at that point. He evaded the issue of whether the versed sine of the *eccentric anomaly* is equivalent to the sum of sines of the *coequated anomaly*.

A more serious difficulty for Kepler's magnetic account of libration was that it required the magnetic poles to maintain a constant inclination to the apsidal line, and there seemed no part of the Earth's body suitable to the test. The Earth's axis, the most likely candidate, did not have the appropriate fixed orientation.

Kepler's inability to provide a sound magnetic theory of libration was not necessarily a failure. As a last resort, he developed an account of how planetary minds could have the knowledge to instantiate the versed sine measure. This account did not require the planets to orient themselves around an empty point in space. On the basis of the position of the fixed stars, the apparent change in the sun's diameter, and its own position, the planetary mind could determine where it has to be. It turns out, Kepler argued, that the sun's apparent diameter changes as the versed sine of the coequated anomaly, and this regularity provides sufficient information for the planet to determine its appropriate position (*AN*, 566; *KGW* III, 359).

The appeal to planetary minds was far from Kepler's first choice, but even so his account of libration avoided the sorts of absurdities he attributed to other models (e.g., having to use empty points as references). Also, as mentioned before, Kepler's geometrical account of libration strikingly unifies the very complex nonuniform motion of the planets, which itself is grounds for confidence, because the production of diverse phenomena from a simple rule indicates its construction on the basis of some archetype. Recall Kepler's belief that because simplicity, unity, and aesthetic elegance are part of God's essence, these features will be reflected in the created world. The versed sine measure has these features, and so Kepler could, on the basis of his archetypes, consider it an accurate depiction of reality.

The Ellipse of Chapter 59

Kepler's success with geometrical unification continued in chapter 59 with his construction of the ellipse. The ellipse can be generated from Kepler's account of libration, and under the ellipse the sum of the areas measured the sum of the distances exactly.[30]

One should note that Kepler used the areas on a circle to measure the sum of distances on an ellipse. He justified this substitution in chapter 59. His goal was to establish that sectors on an ellipse are proportional to the corresponding sectors on a circumscribed circle, and that the areas of the circle are proportional to the corresponding distances of the ellipse. To do this, he inscribed an ellipse in a circle, and dropped perpendiculars from the circumference of the circle to the apsidal line (e.g., KML and EBH in figure 4.10). The ratio of a given perpendicular from the circle to the corresponding one on the ellipse is proportional to the ratio of any other such lines (e.g., KL:ML is proportional to EH:BH). In protheorem II, he argued that the ratio of the area of the circle (AEC) to that of the ellipse (ABC) is proportional to the ratio of the perpendicular from the circle to the corresponding perpendicular from the ellipse (KL:ML). As a result, in protheorem III, the ratio of the area measured by the coequated anomaly in the circle to that in the ellipse (e.g., ANK:ANM) is proportional to the ratio between the length of the corresponding perpendicular from the circle to that of the ellipse (KL and ML). In protheorem IX the ratio of the area of the whole circle to the diametral distances it contains is proportional to the ratio of the area of a given sector (ANK) and the diametral distances that belong to its arc (AK). But these diametral distances (KT, TI) are equal to the corresponding distances of the ellipse (MN, NY), by protheorem XI. As the result of Protheorems IX and XI, by protheorem XII the ratio between the area of a sector on the ellipse sweeping out from the sun (area ANM) and the area of the whole ellipse is proportional to the ratio of the area of a sector of the circle sweeping out from the sun (area ANK) and the area of the whole circle (e.g., AMN:ABC is proportional to AKN:AEC). By these equivalences, the area of the circle is a measure of the sums of distances from the sun when the orbit is elliptical (*AN*, 577–85; *KGW* III, 367–71).

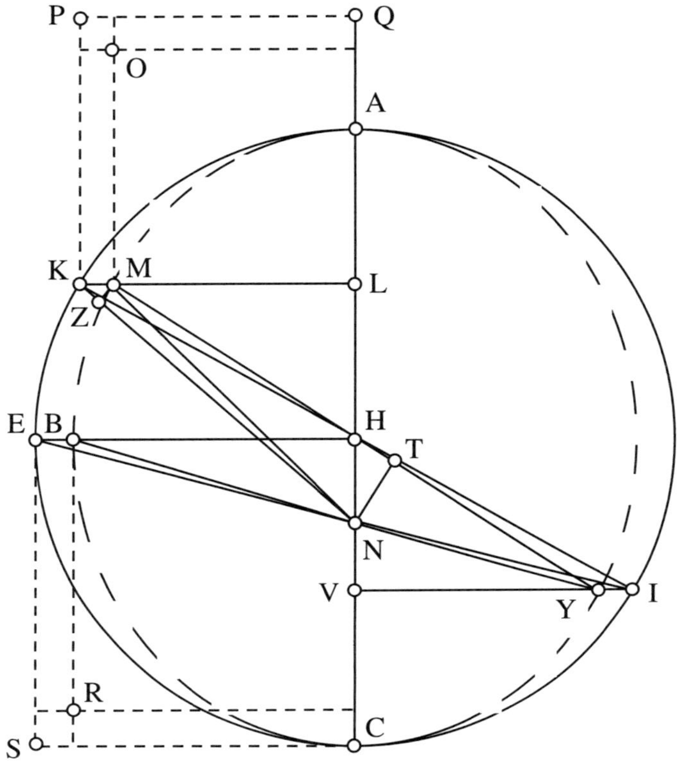

Figure 4.10. Kepler's Ellipse

Donahue sees the emphasis on distances rather than areas in the ellipse as evidence that Kepler was not yet committed to the validity of the area rule (*AN*, 585, n. 16). Another interpretation suggests itself, however. The use of the areas on a circle to measure sums of distances on the ellipse only works because the relationship between the areas and distances on the ellipse is proportional to the relationship between those on the circle. Kepler was connecting elliptical motion to the circle, the former having the virtue of variable motion tractable to a rule, the latter having the virtue of being the most divine plane figures. Demonstrating the relationship between the area rule and the distance law gave Kepler's account the best of both worlds.

One can see why the ellipse was so appealing to Kepler, given its unification of the libration and distance models, the distances and areas of the second law, and its orderly relationship to the circle. This striking convergence was evidence not just for the ellipse, but also for a planned order underlying the universe. I imagine that until this point, Kepler must have felt he was pushing his physical astronomy through by brute force. These results must have buoyed his confidence in being able to find the correct physical account of libration and to complete his archetypal project.

Concluding Remarks: Archetypes in the *Astronomia*

In the *Mysterium* Kepler promoted the use of the a priori method in conjunction with the a posteriori method. Yet we see little of this method explicitly in the *Astronomia*. Additional archetypal hypotheses were not developed here in support of Kepler's major discoveries. He did allude to the Trinity archetype in his arguments for the centrality of the true sun, but this allusion is the only place in the *Astronomia* where the archetypes are markedly prominent. What we have instead is the use of methodological commitments motivated by Kepler's archetypal conception of the universe, and the use of general maxims drawn from his general notion of what a complete archetypal model would look like.

To sum up, Kepler used the Trinity archetype as an a priori argument supporting his a posteriori derivation. The vicarious hypothesis was rejected on methodological grounds presented in the *Mysterium* and *Apologia*, which were themselves motivated by the conception of an archetypally ordered universe. Kepler was able to justify the claim that the sun provides the primary motive force on the grounds that his new physical commitments were philosophically sound. His three-tiered conception of the universe engendered the collapse of the celestial-terrestrial distinction while maintaining the divine-physical distinction. This vision also allowed the ovoid model of chapter 45 to be ruled out and buttressed Kepler's confidence in both his model of libration and the ellipse. He was unable to ground fully in physics his final account of the Martian orbit, but the convergence of various geometrical models made his hypothesis a candidate for archetypal matching. Kepler's new astronomy, though driven by physical considerations, was ultimately grounded in his archetypal commitments to the unification of diverse phenomena and the unifiability of geometrical models.

Possibly the most important function of the archetypes in Kepler's physical astronomy is something I have passed over in silence until now. In Kepler's physical exploration of the Martian orbit, he tended to take two very different approaches. At times he was concerned to account for the mechanism responsible for the orbit. At other times, however, what he considered physical models were not models of the mechanism, but quantitative representations of the motion (e.g., the sine of the coequated anomaly measure of libration). The latter approach was one of the more radical aspects of Kepler's physical astronomy, and to many of Kepler's contemporaries it did not seem methodologically sound. In the next chapter, I place the standard objections to merging the mathematical and physical disciplines in historical context and argue that Kepler's archetypal account removed the conceptual barriers to his quantitative physics.

5

The Aristotelian Kepler

Until now my focus has been the positive influence of Kepler's archetypal cosmology on his epistemology and methodology. Although his cosmology, like his physics, is no longer part of the scientific canon, it retains significance in virtue of its contribution to Kepler's lasting achievements. To the modern mind, the archetypes are valuable for the roles they played in his astronomical work. But it was his contemporaries that Kepler needed to persuade, not us, and it was his physical astronomy that he needed to defend, not his cosmology. Kepler would have found useful a rhetorical platform from which to justify his radical discoveries, and he had the skill to construct one. Both Jardine (1984, 75) and Stephenson (1987, 3) note that Kepler took care to write the *Apologia* and *Astronomia* in a manner designed to win over opposition. I argue that Kepler's archetypal cosmology was a crucial part of this platform, allowing him to avoid straying too far from the Aristotelianism dominant during his time. Despite Kepler's avowedly Platonic and Pythagorean sympathies, his physical astronomy fit well with Aristotle's directives in the *Posterior Analytics*. Perhaps paradoxically, the archetypal cosmology of the *Mysterium* enabled the merging of Platonic and Aristotelian intuitions in Kepler's new astronomy.

The Aristotelian Challenge

The *corpus Aristotelicum*, though no longer the authority it had been in late medieval times, was still the foundation of university curricula in the sixteenth century. This arrangement was true of Tübingen, Kepler's alma mater, although some of Kepler's teachers, notably Andreas Planer and Michael Maestlin, were well acquainted with Neoplatonism and Copernicanism.[1] Schmitt (1975, 489) suggests that despite various challenges to Aristotelian natural philosophy, its overthrow was inhibited by the lack of alternative approaches equally extensive and systematic. As a result, Aristotelianism was taken as the alpha, though by no means the omega, of approaches to natural philosophy.[2]

As noted earlier, in Kepler's time there was cause for concern about the status of astronomy. There were competing astronomical hypotheses, observationally equivalent and seemingly unadjudicable. Moreover, they were incompatible with Aristotle's physical account of the heavens since the best astronomical theories did not restrict themselves to concentric circles.

In the *Posterior Analytics* Aristotle proposed that knowledge is *produced* through the study of essences and is best *presented* as a series of demonstrations whose terms all advert to the same genus and to the essential properties of objects from the perspective of that genus (*Posterior Analytics*, 71b10–25, 73a22–74a3, 75a38–b20). Mathematics and physics were distinct for Aristotle, as the former is the study of form in abstraction from objects, and the latter is concerned with the form and matter of changing objects. As a result, the subject matter, the method of study, and kinds of expected results differed between the two disciplines (*Posterior Analytics* 79a8–12, 81b4; *Physics* 193b22–194b15; *Metaphysics* 1036a4–13). Geometrical astronomy, on this conception, cannot provide knowledge about the actual motions of celestial bodies, because geometry is not about the essential characteristics of moving bodies. By contrast, physics is about the essential characteristics of moving bodies, and so can provide knowledge about such motion. As a result, any conflict between physical and astronomical hypotheses should be ruled in favor of the physical. In addition, because astronomy does not yield knowledge about planetary motion, it lacks the means to adjudicate between competing hypotheses.

Significant Aristotelian elements imbued skeptical attitudes toward astronomy in the fifteenth and sixteenth centuries. Jardine identifies the majority position as the moderate claim that astronomical models are mere calculation devices. The moderate position was skepticism not about physics as a method for obtaining information but about mathematical astronomy. Mathematical and physical treatments of planetary movement were viewed as fundamentally distinct, for decidedly Aristotelian reasons.[3] The following quotation from Benito Pereira (c. 1460–1553) is given at length to show its Aristotelian character and to set the stage for Kepler's breakthrough in merging astronomy and physics.

> Even though the physicist and the astrologer[4] deal with the same heaven, they deal with it in different ways. . . . For my part I think that six differences between physics and astrology can be established. This first difference is this. Physics considers the substance of the heavens and the stars, whether it is ingenerable and incorruptible, whether it is simple or composite, whether it is elementary or is rather a certain fifth essence. Astrology does not consider these matters at all. Secondly, with respect to the heavens the physicist inquires into all the genera of causes, asking for example, whether the heavens have an efficient cause or not, whether they have a material cause, whether the intelligence which moves them is their soul and form, what the final cause of the heaven is, and how it operates. These matters fall outside the concern of the astrologer. Thirdly, of the acci-

> dents of the heavens the astrologer chiefly considers magnitude, shape and motion, insofar as certain mathematical measures are to be found in these accidents, for example, the relations of greater, lesser and equal in distance, nearness and size or proportion. The physicist, however, examining all the accidents of the heavens, even those which are dealt with by the astrologer, considers them indeed in a very different way, as they are derived from the nature of the heavens, as they are in conformity with its substance, as they are necessary to it and for the carrying out of its physical operations, and finally as they are linked with sensible matter. Fourthly, the astrologer is not concerned to seek and posit causes that are true and agree with the nature of things, but only causes of such a kind that he can universally, conveniently and constantly give an account of all those things which appear in the heavens. . . . Fifthly, even though the physicist deals with the same matters as the astrologer, the one demonstrates them *a priori*, whereas the other very often demonstrates them *a posteriori*. . . . Sixthly, wherever it happens that both explain the same thing, the physicist gives the specific and natural causes, whereas the astrologer gives the general and mathematical causes. (quoted in Jardine 1984, 237–38)[5]

Pereira assigned different objects of study, methods, aims, and epistemological interpretations to physics and astronomy on the basis of whether the discipline investigates the essences or merely the accidents of heavenly bodies.

Kepler's response to the problem of observational equivalence was that, although competing astronomical hypotheses might be observationally equivalent, they can be adjudicated on physical (including metaphysical) grounds (*Apologia*, 141–42; *KGW* XX.1, 22). This suggestion was not without precedent. Physical considerations traditionally had been used to constrain astronomical hypotheses. Recall the requirement that planetary motion be resolved into uniform circular motion. What was unusual was Kepler's belief in the possibility of constructing an astronomy both empirically adequate and consistent with physics. As Jardine notes, in Kepler's time many thought the conflict between physics and astronomy resolved by divorcing the two. Empirically adequate astronomy was deemed acceptable, but merely as a predictive tool. Because it did not provide knowledge of the physical causes of motion, it did not need to be consistent with accepted physical principles. The use of physics in astronomy, then, though not prohibited by Aristotle, was ruled out on Aristotelian grounds. Some of Kepler's teachers, notably Maestlin and Planer, followed Aristotle in emphasizing the differences between the mathematical and physical disciplines, and the inappropriateness of mixing the two (Methuen 1998, 188–203). Thus Kepler's approach was in conflict with the Aristotelianism of his day.

Mathematical physical astronomy also conflicted prima facie with Aristotle's *Posterior Analytics*. Kepler wanted to give celestial physics both a mechanistic treatment and a mathematical representation. He made this explicit in a letter to Herwart von Hohenburg (February 10, 1605):

> My goal is to show that the heavenly machine is not a kind of divine living being but similar to a clockwork in so far as almost all the manifold motions are taken care of by one single absolutely simple magnetic bodily force, as in a clockwork all motion is taken care of by a simple weight. And indeed I also show how this physical representation can be presented by calculation and geometrically. (quoted in Caspar [1959] 1993, 136; *KGW* XV, 146).[6]

This goal involved the unconventional mixing of three disciplines: celestial physics, mechanics, and mathematics. Aristotle warned against mixing disciplines together on the grounds that qualities essential to the kind studied by one discipline may be accidental to that studied by another (*Posterior Analytics* 75a27–75b2). Some disciplines could be mixed provided that the principles used were common to both. Aristotle listed optics, astronomy, harmonics, and mechanics as disciplines acceptably mixed with mathematics (*Posterior Analytics* 76a23–25). Mechanical astronomy, however, was not, for it violated Aristotle's celestial-terrestrial distinction, and its mathematical representation violated Aristotle's injunction against the use of mathematics in physics.[7] Mathematics could not yield knowledge in the study of dynamics, according to Aristotle, because quantitative characteristics are not essential to bodies undergoing change.[8] Consequently, it appears that Kepler failed to answer the skeptical challenge to astronomy, at least by Aristotle's standards.

Granted, Kepler avowed allegiance to Plato and Pythagoras, perhaps most clearly in a letter to Galileo (October 13, 1597; *KGW* XIII, 145). So he might not have seen the need to take up the Aristotelian challenge, in either its original form or its Renaissance derivatives. Nevertheless, he still needed to account for the relationship between mathematics and the physical world. Although, following Plato, he believed the world to be structured by eternal archetypes, he also believed the corporeal world to be equally real and knowable. It is important to note that Kepler's views evolved between the *Mysterium* and the *Astronomia*. In the *Mysterium*, he explored empirical regularities but not their causal mechanisms (*MC*, ch. 16, 20; see also Stephenson 1987, 8–20).[9] His conception of the relationship between the archetypal and physical was akin to Plato's conception of the relationship between the master and lower disciplines, where the principles of the various disciplines are inferrable from the Idea of the Good (*Republic*, VI, 511). By the time he wrote the *Astronomia*, Kepler believed that a complete celestial physics would provide an efficient causal account in mechanical terms, and he seemed aware of the implication that one cannot deduce the nature of efficient causes from formal or final causes. As we will see, he avoided drawing physical conclusions from strictly astronomical or archetypal arguments. The physical world was created to instantiate the archetypes, but this fact alone does not determine the material and efficient means of their instantiation. Consequently, one needs to explain how mathematical models can represent physical causation. Kepler's explanation was that the archetypal and the physical are held in strict structural correspondence, and it is in virtue of this correspondence that archetypal considerations can support physical hypotheses. In what follows I outline the role Kepler's archetypes played in his defense of mathematical physics, both as a mixed discipline and as a means of knowing the causes of celestial motion.

Physical Astronomy as a Mixed Discipline

Although it seems sensible to ground astronomy in physics, in Kepler's time such a move was unusual, and in need of defense—a point of which Kepler was well aware. Physics at that time was a qualitative discipline, based on considerations of essential properties of bodies in motion, whereas astronomy was a quantitative, kinematic approach to the movement of celestial bodies. Because physics and astronomy were two vastly different disciplines, it was not obvious that they could be merged. In the *Apologia*, Kepler argued that astronomy is progressive, and that its progress results from combining certain elements of physics with astronomy (Jardine 1984, 222). Kepler wryly commented that Aristotle "is, indeed, to be censured because he mixed with astronomical observations his philosophical reasonings, which were altogether disparate in kind" (*Apologia*, 177; *KGW* XX.1, 43). But Kepler's defense of physical astronomy was not merely a facetious *tu quoque*. His view of the universe as archetypal allowed him to combine astronomy and mathematical physics in a manner that is in fact consistent with Aristotle's conception of natural philosophy as outlined in the *Posterior Analytics*.[10]

In the *Posterior Analytics*, Aristotle allowed for the use of elements from one discipline in another, but only in special circumstances. A deduction may proceed from one discipline to another provided that the first is higher than the second, and the principles used are common to both, where principles are understood to be definitions and basic truths about the objects under study (*Posterior Analytics*, 75b14–21, 76a10–25). At times this move was not merely permitted but recommended by Aristotle. The goal of natural philosophy is not just to show that something is the case but also to explain why. To explain why something is the way it is, one sometimes needs to appeal to a higher discipline (*Posterior Analytics*, 78b35–79a16, 87a32–87b4).[11]

On this view, the highest discipline for Kepler is the archetypal, because it offers final and formal explanations of the construction of the universe. An inference from archetypes to another discipline, then, would be legitimate when structural features of the archetypal domain map the domain of the lower discipline. The use of archetypes in a demonstration of the formal or final features of an object considered under a lower discipline, therefore, would not violate Aristotle's conception of natural philosophy.

Because Kepler saw all aspects of nature as instantiating the archetypes, he conceived of the formal and final features of objects in various disciplines as unified. This conception implies that hypotheses from one discipline can fruitfully bear on those from another, as in the case of using physical hypotheses to help adjudicate between competing astronomical hypotheses. Indeed, since physics provides efficient explanations of astronomical phenomena, physics is a higher discipline than astronomy. Hence it is not only appropriate but perhaps obligatory to ground astronomy in physics.

So far, this approach is consistent with interpreting Kepler as a methodological and metaphysical Platonist. However, I interpret him as a methodological Aristotelian, because while he expected the structural features of each discipline to mirror each other, he still addressed the objects in each discipline with arguments proper to those disciplines. Kepler did not state explicitly in the *Astronomia* that he intended to keep arguments proper to each discipline separate. Instead he spoke of mingling physics and astronomy (*AN*, 47; *KGW* III, 19). Nonetheless, his approach in the *Astronomia* involved keeping physical and astronomical arguments distinct. The "mingling" happened at the level of the conclusions, the structural coherence between the two supporting both.

A quick look at Kepler's outline in the *Astronomia* makes it clear that, for the most part, the astronomical arguments and the physical arguments to the same conclusion appear in different sections. For example, in parts I, IV, and V, he argued "necessarily and geometrically" that the sun is the center of the planetary system. His physical arguments for this conclusion appear in part III.

Even when the physical and astronomical arguments appear in the same section, they are carefully distinguished. For example, though he addressed both the physical and geometrical aspects of his libration theory in chapter 57, he made sure to identify each argument as within its proper domain. Reviewing Kepler's account of libration clarifies his approach. Kepler's first account was that the planet librates as the versed sine of the eccentric anomaly. This account is nice because it unifies the very complicated motion toward and away from the sun. Nonetheless, Kepler was dissatisfied, for the vertex of the eccentric anomaly is at an empty point in space; and because geometrical points cannot be causes, this calculation could not count as a measurement of the strength of the physical force exerted by the sun on the magnetic poles of the planet. He needed a measurement that related the sun's body to the planet's magnetic poles. His second account—that the planet librates as the sum of sines of the co-equated anomaly—accomplished this task.

Nonetheless, Kepler was not satisfied with this result, because he could not develop a satisfactory qualitative account of how the sun's magnetic force interacted with the planet's magnetic poles. Nothing on the Earth has the required orientation to function as these poles. Immediately after observing this problem he wrote: "Now that we have obtained from the observations the laws and quantitative characteristics of this reciprocation [libration] by which the sun's apparent diameter is varied . . . it now remains for us to see whether those laws might plausibly be known by the planets" (*AN*, 560; *KGW* III, 356). After writing this statement, he continued his investigation into the possible causal mechanisms of planetary libration. Clearly, for Kepler a physical description of the causes of motion required a qualitative account of the mechanism producing the effect. Kepler took his second account of libration to count as a possible measure, but not as a physical theory, because it is insufficiently grounded in qualitative theory. While it must have been tempting to conclude that he had succeeded in grounding his libratory account in physics (this was, after all, his goal), his restraint suggests a concern with keeping astronomical and physical arguments separate.

One wonders why Kepler spoke of mingling physics and astronomy when he was so careful to keep his physical and astronomical arguments apart. I do not have an easy answer, only a speculation. Given the tension between Aristotle's physics and Ptolemaic astronomy, and the resultant separation of the two disciplines, requiring consistency between astronomical and physical conclusions might have seemed a radical sort of mingling. Later, Kepler spoke more carefully on the matter, perhaps in response to criticisms of the mingling. One imagines Kepler keenly provoked by Maestlin's remark that physical astronomy did not restrict itself to the proper principles of astronomy. Perhaps it was on the basis of considering such criticism that Kepler wrote, in the *Epitome*:

> Astronomers should not be granted excessive license to conceive anything they please without reason: on the contrary, it is also necessary for you to establish the probable causes of your Hypotheses which you recommend as the true causes of Appearances. Hence, you must first establish the principles of your Astronomy in a higher science, namely Physics or Metaphysics; although you must not neglect those Geometrical, Physical and Metaphysical arguments which are supplied to you by the very separation of the disciplines proper, in addition to those matters pertaining to the higher disciplines, provided you mingle no begging of the Principle. As a result of the method, the Astronomer (master of his proposition to the extent that he has devised the causes of movements, causes which are

> in accord with reason [*rationi consentaneas*], and are fit to effect everything in the History of Observations) now may put into single aspect what he had considered part by part before. (quoted in Westman 1972, 261–62; *KGW* VII, 25)[12]

Here, while Kepler reiterated the importance of grounding astronomy in a higher science (a move consistent with Aristotelianism), he also emphasized that each discipline be considered separately. Only then could the parts be put together. This respect for disciplinary boundaries is very Aristotelian. Indeed, Kepler's Platonic sympathies fail to explain it. On the basis of this explicit statement, and its consistency with Kepler's approach in the *Astronomia*, I submit that the method avowed in the *Epitome* was the one used in the *Astronomia*.

Notice that, except for his commitment to grounding astronomy in physics and metaphysics, Kepler's attitude is similar to Pereira's. The separate consideration of arguments belonging to each discipline accords with Pereira's view. Physics and metaphysics still determine the nature of the heavens and the causes of the motions, and they do so "a priori" (recall Kepler's discussion of the status of physics in chapter 47 of the *Astronomia*). So while physics, metaphysics, and astronomy are still distinct disciplines, they must be consistent with each other. Given Kepler's belief that the universe is, for the most part, accessible to the human mind, the skeptical resolution of the conflict between physics and astronomy should be viewed as temporary, to be replaced when one has better theories available. One should not be satisfied until physics and astronomy are consistent.

Aristotle's physics, however, was inadequate to the task, not only because of its incompatibility with the heliocentric hypothesis, but also because it was implausible on Aristotle's own terms. Kepler argued that one implication of Aristotelian physics is that geometrical points are physical causes. This claim is illustrated by Aristotle's view that the heavy is attracted to the center of the universe qua center (*AN*, 54–55; *KGW* III, 24–25). Not only is it implausible to suppose that geometrical points can attract or repel matter, or serve as markers by which matter can orient itself, but, as Kepler was well aware, Aristotle denied mathematical objects causal efficacy.[13]

The Mathematical and the Physical

Given the unification of disciplines under the archetypes, Kepler could apply terrestrial physics to celestial bodies.[14] In viewing physics as unified in this way, Kepler drew on the tradition of mechanics, the theoretical and practical study of machines. In Kepler's time it was plausible to consider this discipline an acceptable mixing of mathematics and physics.[15]

Kepler's use of analogy is telling of his acceptance of the use of mathematics in physics. He frequently drew analogies between celestial causes and the action of levers, or the behavior of light, both of which yielded to mathematical demonstrations. Even his term for planetary motion toward and away from the sun—libration—is borrowed from the term in mechanics for the motion of a lever or balance (*AN*, 556; *KGW* III, 353). In chapter 57 of the *Astronomia*, Kepler reduced libratory motion to the action of a lever, on the assumption that natural causes are subject to the law of the lever: "And since the strength of this angle" is natural,[16] it will follow the same ratio as the balance" (*AN*, 556; *KGW* III, 353). Planetary libration, then, when conceived as the product of magnetic forces, "approaches and recedes according to the law of the lever" (*AN*, 559; *KGW* III, 355). By connecting celestial physics with mechanics, Kepler tapped into a source of previously accepted mathematical representations of physical bodies.

Those who viewed mechanics as mathematically tractable, as Aristotle and Kepler did, needed to explain this tractability by elucidating the relationship between mathematics and the physical world. Harper and Smith note that since "[m]echanics was the enterprise of finding *mathematical* principles that describe motions and other phenomenal regularities *abstractly*" (Harper and Smith 1995, 124), alternative principles could be consistent with the phenomena. An account of the relationship between mathematics and the physical world would provide the appropriate decision procedure. For Aristotle, mathematical and physical approaches are related by their applicability to the same objects, although their perspectives are very different. The mathematician studies the form of material objects in abstraction, whereas the physicist studies material objects qualitatively (*Physics* 193b23–194a5). Leibler, one of Kepler's teachers, also held this view (Methuen, 1998, 194). Likewise for Kepler, the quantitative exists only as an idea unless it is the form of an object:

> If we know the definition of matter, I think it will be fairly clear why God created matter and not any other thing in the beginning. I say that what God intended was quantity. To achieve it he needed everything which pertains to the essence of matter; and quantity is a form of matter, in virtue of its being matter, and the source of its definition. (*MC*, 93; *KGW* I, 23).[17]

If we bear in mind Kepler's archetypal cosmology, the relationship between mathematics and the physical world is that the *correct* mathematical description of a body matches an archetypal model in its material form. Kepler made this relationship explicit in the *Harmonice*, in which he classified the kinds of mathematical relations that could participate in God's creative plan (bks. I–II). For both Kepler and Aristotle, by itself the mathematical perspective cannot reveal information about physical causes.[18]

When grounded in physical theory, however, mathematical demonstrations of physical causes are acceptable.

For the most part Kepler's approach to mechanics in the *Astronomia* was Aristotelian, that is, qualitative. In chapter 33 he introduced his arguments to the sun as the primary motive force with a discussion of qualitative principles. For example, when he argued that the cause of motion must reside in the body of either the sun or the planet, he relied on a conception of the essential nature of distance: "And since distance is a kind of relation whose essence resides in end points, while of relation itself, without respect to end points, there can be no efficient cause, it therefore follows (as has been said) that the cause of the variation of intensity of motion inheres in one or the other of the end points" (*AN*, 377; *KGW* III, 236). That efficient causes cannot be accounted for without reference to essential characteristics coheres with Aristotle's intuitions about how one approaches explanation. The introductory discussion of magnetic properties in chapters 34 and 35 is qualitative in nature. Of interest to Kepler at this point was the comparison between the action of a magnet and that of the sun, and so he did not appeal here to mathematical demonstrations. In chapter 39 his arguments against the possibility of epicyclic motion were based on qualitative axioms, in particular the axiom that only bodies can exude forces. Even in chapter 57, where Kepler represented libratory motion quantitatively, he attempted to show how the mathematical account could be derived from the qualities of magnets.

These qualitative principles, though non-Aristotelian in content, do not offend Aristotle's conception of natural philosophy.[19] Consider the following two principles. Kepler constructed physical models on the assumption that only physical bodies can exert forces. He rejected the mean sun as a basis for planetary models, and the epicycle model as anything but a mathematical tool, for these require movement around a mathematical point. He also believed that forces were subject to physical constraints. As discussed earlier, Kepler thought that when a force spreads out over a greater area, its effect is weakened, and so the planet's delay is proportional to its distance from the sun. Both of these principles, under Aristotle's framework, can be construed as belonging to the same genus under the discipline of physics.

Although Kepler began with qualitative principles, his mathematical models of physical motion were more successful. The attempt to show the magnetic character of the force that moved the planets was ultimately unsuccessful. In chapter 57, after a lengthy attempt to compare magnetic properties and planetary motion, Kepler claimed to have shown only "the general possibility of the proposed mechanism" (*AN*, 559; *KGW* III, 355). By that time, what Kepler had was an imperfect analogy (the magnet) and an elegant mathematical model (the versed sine measure) of planetary motion. Although his qualitative theory fell short, he had a route open to him that Aristotle did not, a route that has not, I believe, been brought fully to light. Given Aristotle's cosmology, mathematical demonstrations can be used in mechanics only if based on solid qualitative theory. On its own, mathematics is indifferent to physical change and hence cannot provide information about physical motion. In contrast, given his cosmology, Kepler had the option of attaching mathematics to either a qualitative *or* an archetypal theory. Unlike the mathematical, the archetypal is not indifferent to physical change for the physical was created to mirror the archetypal. As a result, where physical considerations fail, archetypal considerations can fill the gap, revealing the formal aspects of physical bodies undergoing change.

It is important to grasp fully Kepler's view of the relationship between the archetypal, the mathematical, and the physical. Even though archetypes do not have the power to cause physical change, such change is effected for the purpose of harmonizing with the archetypal: "[Archetypes] are the cause of natural things . . . [though] I confess that they would have possessed no force, if God himself had not had regard to them in the act of Creation" (*MC*, 125; *KGW* VIII, 62).[20] As a result, though archetypally significant mathematical models will not account for efficient causes, they *will* differentiate spurious and causal regularities, which in turn will help decide between various mathematical descriptions of the same phenomena. Although this method would not be fully satisfactory to a realist of Kepler's stripe, it would speak to the plausibility of the physical intuitions on which such models are based.

Several features of a hypothesis jointly indicate whether it corresponds to archetypes. It must have aesthetic appeal and be expressible geometrically (recall Kepler's rejection in the *Astronomia* of the ovoid model of chapter 45). It must also conform to structural aspects found elsewhere in nature, because nature "makes use of the fewest possible means" (*AN*, 51; *KGW* III, 22), in the sense that comparatively few archetypes form the basis for all natural phenomena.

It is my contention that attaching a mathematical model to an archetypal model was Kepler's purpose in proposing his second account of libration. Although he called this account "physical," he clearly did not intend it as a qualitative physical theory. Moreover, he knew that his qualitative theory was problematic, as he could not account for direction of the magnetic poles on the Earth. Instead he was showing that his mathematical description of libration was equivalent to the mathematical description of some two-magnet system, one unlikely to be a description of the actual cause. Kepler clarified this view more in a letter to Fabricius (October 1605). After pointing out the problem with the Earth's axis, he stated that he had at least demonstrated the mathematical equivalence of the two-magnet system and planetary libration, even though the two systems are of seemingly different kinds.[21]

This comment counts as a response to the problem if Kepler is read as arguing that the common mathematical description is evidence for the *analogy* between the two systems. His emphasis that they are different in kind supports this interpretation. Because God chose to replicate pleasing patterns, finding the same pattern in different disciplines is evidence for the archetypal nature of that pattern. This explains why Kepler provided examples of other kinds of physical systems that would produce the same pattern, the two-magnet system, and the boat with turning oar (*AN*, 549; *KGW* III, 349). Clearly the boat example was intended to illuminate not the kind of force at work, but the pattern produced, for Kepler did not believe the two librating forces to be similar.[22] Even in the magnet case, he was not arguing to the efficient cause of the motion. His point was that the two systems could be described *precisely* by the same mathematical description, and this precision gave him confidence that he had uncovered an archetypal pattern. He took attachment to an archetypal pattern to suggest causal rather than accidental patterning. While a mathematical model so attached is not a physical account of the cause, it is plausibly correct and suggests that such an account can be formulated. Thus, Kepler's failure to provide a satisfactory qualitative account is merely an indication of the need for further research, rather than a serious problem for his account.

At this point, however, Kepler's archetypal justification fell short of ideal. Ideally, a full explanation includes an astronomical, a physical, and an archetypal account, all of which cohere to provide converging mutual support. But an archetypal explanation of planetary libration had to wait until he wrote the *Harmonice*. At best what he had was a libration theory indicating, by its aesthetic features, possible conformity with an archetypal model.

The libration theory Kepler developed in the *Astronomia* was strictly astronomical, though its character intimated both physical and archetypal significance. Despite this significance, he refrained from drawing physical or archetypal conclusions, which suggests an effort to avoid illegitimate crossings of disciplinary boundaries.

Viewed in this way, Kepler's mathematical physics does not violate Aristotle's conception of appropriate disciplinary boundaries in the *Posterior Analytics*, though it does violate the Aristotelianism (specifically the mathematical astronomy-physics distinction) prevalent in Kepler's time. The inferences made are from the structure of objects in one domain to the structure of those in another. The inferences are not cross-genus, because in Kepler's archetypal vision the structural properties of all physical objects share principles with the archetypal domain. In the case of libration theory, where he did not yet have an archetypal account, Kepler restricted himself to commenting that such an account was plausibly forthcoming. Were the inference from the structure of archetypes, or from archetypally significant astronomical models, to an account of efficient causes, this approach would be anti-Aristotelian, and decidedly so.

Concluding Remarks

I hope to have made clear one of the ways in which Kepler's archetypal views both motivated and justified his physical astronomy. I also hope to have made clear that the notion of structural mirroring between the archetypal and the physical allowed Kepler to conduct his physical astronomy in accordance with Aristotle's conception of natural philosophy. On this interpretation Kepler was not being disingenuous when, on the title page of book IV of the *Epitome*, he referred to book IV as a supplement to Aristotle's *De caelo*, and compared his approach with Aristotle's. Kepler could justify his innovative hypotheses within a conservative Aristotelian framework.

Kepler's archetypes were more than mere window dressing on Copernicanism, more than a fanciful project to distract him from personal troubles (he wrote the *Harmonice* after his daughter died, and while his mother was being tried for witchcraft). His cosmology supported methodological innovations and justified them as truth-linked, it delimited the kinds of hypotheses proper to physical astronomy, it provided converging support for both astronomical and physical hypotheses, and it mapped the mathematical onto the physical. This interpretation should highlight both the importance Kepler attached to completing his archetypal project and its significance for our understanding of Kepler's achievements. It is with this in mind that I turn next to a discussion of Kepler's *Harmonice*.

6

The *Harmonice mundi*

Kepler and Ptolemy

Kepler's imagination was captured in 1599 by the idea that planetary motion could be explained by harmonic theory. He wrote letters to Edward Bruce (who was associated with Galileo), Herwart von Hohenburg, and Michael Maestlin about a harmonic theory that predicted planetary delays better than the nascent physical theory in chapter 20 of the *Mysterium* (*KGW* XIV 7–16, 21–41, 43–59).[1] He also requested that von Hohenburg send him a copy of Ptolemy's *Harmonics*, though he found it to be a frustratingly bad Latin translation. He did not receive the *Harmonics* in its original Greek until 1607 (Stephenson 1994, 37). He was struck by the similarity between his thought and Ptolemy's, despite being separated from Ptolemy by both astronomical theory and many centuries:

> [M]y appetite was particularly intensified and my purpose stimulated by the reading of the *Harmony* of Ptolemy. . . . There I found unexpectedly, and to my great wonder, that almost the whole of his third book was given up to the same study of the celestial harmony, one thousand five hundred years before. Yet at that period much was still lacking in astronomy, and Ptolemy, by an unfortunate attempt, might have brought others to despair. . . . he seemed to have related some pleasant Pythagorean dream rather than to have assisted philosophy; but I was emphatically strengthened in the pursuit of my purpose, on the one hand by the crudity of the ancient astronomy, and on the other by the actual agreement, precise in every detail, of both views, after an interval of fifteen centuries. . . . This identity of conception, on the conformation of the world, in the minds of two men who had given themselves wholly to the study of nature, was the finger of God, to borrow the Hebrew phrase, since neither had guided the other to tread this path. (*HM*, 390–91; *KGW* VI, 289)

Kepler and Ptolemy agreed that harmonic ratios could be found in both the heavens and human soul. Ptolemy believed that the soul's harmonic ratios explained its sensitivity to music and astrological influences. Indeed, he recommended the study of astronomy as a means of understanding and improving the human soul (Taub 1993).

Kepler mentioned his outline for his study of world harmonies in a letter to von Hohenburg (December 14, 1599; *KGW* XIV, 100):

1. Geometrical, on constructible figures.
2. Arithmetical, on solid ratios.
3. Musical, on the causes of harmonies.
4. Astrological, on the causes of Aspects.
5. Astronomical, on the causes of the periodic motions.

(quoted in Field 1988, 97) Although this plan substantially resembles the final table of contents, Kepler did not begin writing the book in earnest until much later (1617). He finished the *Harmonice* (*KGW* VI) on May 27, 1618, but discovered the third law almost a year later (May 15, 1619). To include this important finding in the published version, Kepler very quickly had to revise book V, with the typesetting already well underway (Stephenson 1994, 125).

In the published version, the first book is devoted to showing how the regular solids are constructible from the circle, and how their constructibility determines their rank ordering. Book II addresses the manner in which regular figures can be combined to form either plane figures (tessellations) or figures whose vertices are all of the same type. If figures can be combined in this manner, then they are congruent. When the plane figures are the same shape, the highest level of congruence is expressed in the regular polyhedra. Kepler constructed both polygons and polyhedra in this manner, and offered two new regular polyhedra he had discovered (figure 6.1, Ss and Tt; from *HM*, 111). These formations are also rank ordered, but as Kepler observed, the rank ordering in book I does not correspond to that in book II. These results, particularly those from book I, are used in book III to show how musical harmonies can be explained by geometrical figures. Kepler also linked musical harmonies to human emotions, and commented on the implications of mathematics for politics. Book IV links astrology to mathematical harmony, and includes a discussion of the significance of mathematics to politics. This latter section includes, as does book V, a study of how God created the social and natural order to be in tune with mathematical harmonies. Here I am concerned primarily with book V, which focuses on the archetypal causes of the periodic motions and eccentricities.

My focus, as in preceding chapters, is Kepler's use of archetypes. The stated purpose of book V is to explain by appeal to harmonic archetypes "the origin . . . of the Eccentricities, Semidiameters, and the Periodic Times" (*HM*, 387; *KGW* VI, 287). Here, as elsewhere, Kepler's goal was to provide a formal and final explanation of things. The corresponding efficient causes that he had been unable to account for fully in the *Astronomia* he would account for here as resulting from the archetypally justified distances and eccentricities.

In setting the stage for the *Harmonice*, I discuss how Kepler followed the Pythagoreans and Ptolemy in translating harmonies into mathematical

Figure 6.1. Kepler's New Polyhedra

ratios. Next, I focus on Kepler's concept of knowability in books I–IV, in which he argued explicitly that through mathematics the human mind can access the divine. This chapter begins with a synopsis of book V.[2] Then I discuss the relationship between Kepler's mathematical and harmonic accounts of planetary libration, and how the latter was intended to support the former. It turns out that Kepler's view of the cosmological significance of geometry shifted in response to difficulties he found in constructing planetary harmonies. As I hope to make clear, Kepler was unwilling to abandon the principle that everything was created for a reason, which led him to modify the cosmology that had aided his construction of the new astronomy. He did not flag these changes, nor did he explicitly discuss their influence on the rest of his work, and so in closing I speculate on the possible effects this shift had on Kepler's view of archetypal and physical astronomy.

In the *Harmonice*, more pronouncedly than in the *Mysterium* or *Astronomia*, Kepler announced his discoveries with exuberant delight. One might take this to indicate that the *Harmonice* accomplished Kepler's archetypal task, and in a decidedly precise and elegant manner. On both cursory and careful readings, however, Kepler's ecstatic judgment appears unwarranted. The construction of harmonic archetypes is not only complex but imprecise. The ratios used are often just shy of the harmonic, and justifications for such divergences abound. Gingerich comments that, "[f]rom anyone else, the carefully crafted excuses and scales would be considered the edifice of a madman" (Gingerich 1992, 60). Kepler himself wondered, though not for very long, "whether the highest creative wisdom would have been taken up with searching out these thin little arguments?" (*HM*, 466; *KGW* VI, 342). For Kepler, the complexity of the harmonic archetypes was far less absurd than the notion that God created the universe, exactly as it is, without a reason.

Before turning to the *Harmonice*, I want to summarize Kepler's method of relating mathematical ratios to harmonic concordances. Like Ptolemy and the Pythagoreans, Kepler grounded harmonics in mathematics, where the difference between melodic and unmelodic sequences is that the melodic can be translated into mathematically significant ratios. The Pythagoreans used the *kanōn*, or monochord (a string stretched across a movable bridge), as a kind of translation manual between the heard music and the mathematics underlying it. The bridge was moved to divide the string in different proportions, recorded as ratios, of its full length. For example, the interval between the note plucked at full length and another plucked at half length is one octave. The octave is then represented by the ratio 2:1. Similarly, the fifth is represented as 3:2, the fourth as 4:3, the ditone as 5:4 and the tone as 9:8 (Barker 1989b, 291–93).

There are several differences between Kepler's treatment of mathematical harmonies and those of Ptolemy and the Pythagoreans. Kepler grounded harmony in geometry rather than arithmetic. Ptolemy and the Pythagoreans argued that the difference between concordant and discordant ratios lies in the arithmetical significance of the ratios. For this reason some gave special status to multiples ($mn{:}n$) or epimoric ratios ($n{+}1{:}n$) (Barker 1989a, 272–73). Ptolemy conceived of the concordant and discordant not as distinct types but on a continuum. Ratios were more or less concordant, in virtue of their distance from the ratio 2:1 (Barker 1989a, 273).

Another important difference is Kepler's advocacy of the just intonation system, which he refers to as Ptolemy's system. Both the Pythagorean and just intonation systems are based on mathematical principles. Only just intonation, however, permits the third and sixth as consonants, both of which were significant to polyphonic music in Kepler's time (Walker [1967] 1978). On the Pythagoreans, Kepler noted that "They advanced to a certain point by the judgement of their ears, but soon abandoning their leadership completed the rest of the journey by following erroneous Reason, so to speak dragging their ears astray by force, and ordering them outright to turn deaf" (Gingerich 1992, 48; *KGW* VI, 120). Although musical ratios had to be empirically adequate, they also had to be mathematically constructed, and on these grounds Kepler rejected the tempered scale defended by Vincenzo Galilei, Galileo's father. Kepler found the tempered scale, though aesthetically pleasing, mathematically problematic (Gingerich 1992, 57).

Debates on the place of reason and the senses in music theory have roots in ancient music theory. Whereas the Pythagoreans based music theory on mathematics, Aristoxenus vigorously disagreed, basing it solely on judgments of the ear. Ptolemy used both, for he thought reason best for constructing melodic sequences, and perception best for judging them (Barker 1989b, 276–78). Likewise for Kepler, music theory, like astronomical theory, required converging support from both the empirical and the theoretical.

Yet another difference between the music theory of antiquity and that of Kepler's time is in their respective attitudes to polyphonic harmony. The Pythagoreans, Ptolemy, and Aristoxenus held that concordance is the relationship between notes sounded sequentially (Barker 1989a, 14–15). But polyphonic music had become popular by Kepler's time, and he used polyphonic harmony in his construction of the planetary voices.

Despite the similarities between Kepler and Ptolemy, their differences in applying harmonies to planetary theory cannot be attributed solely to their different interpretations of harmony and different identifications of the central body of the heavens. Ptolemy compared astronomical data with music theory without great precision. He did not construct a harmonic theory for the purpose of generating astronomical predictions. His only concern was general similarities between the two.

Ptolemy began the latter half of book III of the *Harmonics* with a look at the standard scalar system, the complete *systēma*. The complete *systēma* is a double octave with fifteen notes, grouped into tetrachords, which are progressions spanning a fourth. The notes are named according to either their position from lowest to highest (*thesis*), or their function relative to other notes (*dynamis*). This double octave can be rearranged to exhibit different *tonoi* (named after the ethnic groups with whom they were associated), modulations that effectively move intervals of the double octave

TABLE 6.1
The Greater Perfect System

Tetrachords	*Notes*
	Nētē hyperbolaeōn
Tetrachord	Paranētē hyperbolaeōn
hyperbolaeōn	Tritē hyperbolaeōn
	Nētē diezeugmenōn
Tetrachord	Paranētē diezeugmenōn
diezeugmenōn	Tritē diezeugmenōn
	Paramesē
	Mesē
Tetrachord	Lichanos mesōn
mesōn	Parhypatē mesōn
	Hypatē mesōn
Tetrachord	Licanos hypatōn
hypatōn	Parhypatē hypatōn
	Hypatē hypatōn
	Proslambanomenos

Note: The top two tetrachords, *hyperbolaeōn* and *diezeugmenōn*, are conjunct (i.e., they share the note called *nētē diezeugmenōn*). The middle tetrachords, *diezeugmenōn* and *mesōn*, are disjunct (they do not share a note). The lower two tetrachords, *mesōn* and *hypatōn*, are conjunct, sharing the note *hypatē mesōn*. A final note, *proslambanomenos*, is added to complete the double octave. The top and bottom notes of each tetrachord are fixed. The middle two notes are movable, determining whether the genus of scale is diatonic, enharmonic, or chromatic (a classification based on the distance between each note in the scale).

systēma from top to bottom. Consider the standard *systēma* in table 6.1 (Greater Perfect System) with notes named according to *thesis* (Barker 1989b, 284–85, 288–89).

When modulating *tonoi*, the top note of the Greater Perfect System (*Nētē hyperbolaeōn*) is treated as identical to the bottom note (*proslambanomenos*). Like the motions of the heavenly bodies, the notes as defined by function [*dynamis*] have no fixed starting point, and move in a circuit. If the Greater Perfect System is conceived as a circle with *nētē hyperbolaeōn* joined with *proslambanomenos*, significant points on the zodiac correspond to significant parts of the Greater Perfect System. The equinoctial point corresponds to *mesē*. In astrological terms, the octave is then associated with opposition. The twelve parts of the zodiac are associated with the twelve tones of the double octave.[3] The remaining concords are similarly mapped onto the zodiac circle (Barker 1989b, 380–83).

Ptolemy divided the motions of stars into three categories: longitudinal motions (risings and settings), vertical motions (further from and closer to the Earth), and lateral motions (shifting either north or south). He compared ascending and descending the musical scale to the rise and setting of stars. He compared the vertical motions to the *genera*, the *genus* determined by where the tetrachord's two internal notes are placed. He likened the moderate chromatic to motions at intermediate distances and velocities, the enharmonic *genus* to the slowest velocities, the diatonic to the fastest velocities, and the modulations of the *tonoi* to the lateral motions (Barker 1989b, 384–88).

Chapters 14 and 15 of the *Harmonics* were lost (Barker 1989b, 390), although the titles remain, indicating that in those chapters Ptolemy discussed the relationship between harmonic ratios and planetary movements.[4] What remains of the *Harmonics* on this topic is general, and closed to the kind of testing Kepler required of his own harmonic theory.

The Essence of Archetypal Mathematics

In the introduction to book I of the *Harmonice*, Kepler claimed that a study of harmony should begin with geometry: "We must seek the causes of the harmonic proportions in the divisions of a circle into equal aliquot parts, which are made geometrically and knowably, that is, from the constructible regular plane figures" (*HM*, 9; *KGW* VI, 15). The reason why harmony should be so grounded in geometry derives from the relationship between knowable figures and the archetypes. Kepler defined "knowability" as follows:

> **VIII Definition**
>
> A quantity is said to be knowable if it is either itself immediately measurable by the diameter, if it is a line; or by its [the diameter's] square if a surface: or the quantity in question is at least formed from quantities such that by some definite geometrical connection, in some series [of operations] however long, they at last depend upon the diameter or its square. (*HM*, 19; KGW VI, 21–22)

That is, a quantity must be derivable from the diameter (or its square) of a circle. Definitions XII to XV classify the construction of figures into degrees of knowability, where knowability decreases as the complexity of deduction, or "distance" from the circle increases.

This concept of knowability allows Kepler to distinguish between physically plausible and implausible mathematical descriptions. For Kepler, unknowable figures could not have played a role in creation: "For if they are knowable, then they can enter the Mind and into the shaping of the archetype; but if they are unknowable (in the sense which has been explained in Book I) then they have remained outside the Mind of the eternal Craftsman, and have in no way matched the archetype" (*HM* 139; *KGW* VI, 100–101). We finally get from Kepler an explicit classification of the kinds of geometrical and harmonic constructions that are likely to have archetypal and, thus, physical significance. Knowability places limits on the kinds of things that can exist, for only knowable constructions could have been in God's mind to create.

Despite the general acceptance of using mathematics in harmonic theory, Kepler argued for the legitimacy of this practice. He first located harmony not in the sounds themselves, but in relations between sounds. Because relational harmony is ordered, and order is mathematical, the use of mathematics in harmonics is legitimate. Furthermore, harmony and quantity cannot exist without sound and matter respectively, or apart from a perceiving soul, and therefore are similar kinds of thing (*HM*, 290–91; *KGW* VI, 212). Here Kepler was following Aristotle's requirement in the *Posterior Analytics* that mixed sciences have principles in common.

In explaining how the soul recognizes harmony and quantity, Kepler sided with Proclus and Plato against Aristotle, appealing to the archetypes. Plato proposed in the *Meno* that the human mind is imprinted with the Forms and learns of them as a kind of recollection. Aristotle, by contrast, claimed that the mind is "just a blank sheet, so that nothing is written on it, not even any mathematicals" (*HM*, 298; *KGW* VI, 218). In other words, mathematicals are not recalled by an already knowing soul, but abstracted by the mind from physical objects. Following Proclus, Kepler disagreed vehemently, referring to Aristotle's argument as stupid (*HM*, 298; *KGW* VI, 218). He did this on the grounds that without an archetype with which to compare them, the soul would be unable to recognize quantities and harmonies (*HM*, 299; *KGW* VI, 219).[5] Kepler agreed, however, that material quantities, having particular shapes and sizes cannot exist apart from matter and must be abstracted from it (*HM*, 290–91; *KGW* VI, 212).

It is in virtue of being imprinted with God's image that our minds contain ideas of quantity and harmony:

> Geometry, which before the origin of things was coeternal with the divine mind and is God himself (for what could there be in God which would not be God himself?), supplied God with patterns for the creation of the world, and passed over to Man along with the image of God; and was not in fact taken in through the eyes. (*HM*, 304; *KGW* VI, 223)

The implication is that these ideas aid study of the physical world. As argued in the previous chapter, this relationship should not be taken to mean that mathematics can tell us everything we need to know about physics. It only indicates the structure of physical systems. Kepler did not discuss this matter explicitly in the *Harmonice* as he did in the *Epitome*, but he did hint at it: "For shapes are in the archetype prior to their being in the product, in the divine mind prior to being in creatures, *differently indeed in respect of their subject, but the same in the form of their essence*" (*HM*, 9–10, emphasis added; *KGW* VI, 15).

As we can see, Kepler continued to believe that the world is both archetypally structured and knowable through the archetypes. He added to this the notion that knowability requires geometrical constructibility. This requirement explains his uneasiness with astronomical models that could not be constructed properly, such as the ovoid model of Chapter 45 of the *Astronomia*.

Synopsis of Book V

The *Harmonice* is a complex weave, and I only follow a few of its threads, focusing on Kepler's archetypal determination of the extreme motions, eccentricities, and distances. Little attention will be paid to the *genera* of the planets, the completeness of the chords they sound, and so on. Chapters 1–4 and 9 are the most important for my purposes, although 5–8 will be mentioned briefly. My purpose in maintaining a narrow focus is to present as clear and uncluttered a picture as possible of Kepler's archetypal derivation of astronomy. Bear in mind, however, that this picture cannot help but be somewhat distorted. Part of Kepler's justification for his archetypal system is the way all the parts hang together.

Kepler intended to develop in book V a system of planetary harmonies that explains the periods and eccentricities of the planets.[6] To do this properly, in accordance with his cosmological commitment to the primacy of geometry, he sought to base planetary harmonies on the nested-sphere model. Although he had already given a geometrical account of the distinction between the concordant and discordant in book III, the explicit relationship between geometric and harmonic archetypes was not yet established. He devoted the first two chapters to this end.

The first chapter is drawn from the polyhedral hypothesis of the *Mysterium*. Kepler knew that the polyhedral hypothesis was inadequate, but still considered it archetypal. In linking the polyhedra to harmonic consonances, his first step was to construct two "weddings," males (the cube and dodecahedron) to females (the octahedron and icosahedron). The tetrahedron was considered hermaphroditic (*HM*, 397; *KGW* VI, 292). Kepler discussed the importance of the dodecahedron-icosahedron wedding in chapter 9, arguing that despite its difficult inclusion in the universal harmonies, it nonetheless plays a critical role in determining the *genera* of the song.

Chapter 2 involves relating the polyhedra to harmonic ratios. Kepler focused on the ratios between adjacent spheres, this being "more appropriate to the present work" (*HM*, 401; *KGW* VI, 295) than other methods. Relating harmony to geometry is important because, as later stated in Chapter 9, "the proportions of the orbits, and their individual eccentricities, issued simultaneously from the archetype" (*HM*, 451; *KGW* VI, 330).

Kepler assigned the ratios 1:2 and 3:5 to the cube-octahedron wedding, as these are the closest harmonic ratios to the square root of 1:3, the ratio between the spheres of the cube and octahedron. For similar reasons he assigned 3:4, 5:8, 4:5, and 5:6 to the dodecahedron-cosahedron wedding, and 1:8 and 1:4 to the tetrahedron, notwithstanding the 1:3 ratio of its surrounding spheres. Wishing to leave 1:3 for the cube, Kepler argued that since the ratio for tetrahedral spheres is the square of those for cube-octahedron, tetrahedral harmonies should be the square of cubic harmonies; that is, 1:4 and 1:9. Because the ratio 1:9 is not harmonic, 1:8 would do (*HM*, 402; *KGW* VI, 296).

The assignment of these particular ratios is clearly a result of Kepler's desire to merge, by any means necessary, his harmonic and polyhedral hypothesis. This forced merger cannot help but cause wonder, for his earlier work was much more rigorous and precise. Even in the *Mysterium* where Kepler's a priori ordering of solids appears ad hoc, there is still a comfortable fit between the model he constructed and his reasons for so constructing it.

Chapter 3 summarizes, in thirteen propositions, the astronomical background to Kepler's harmonic theory. Propositions 1–4 concern the order and spacing of planetary orbits. The ordering is Copernican, the spacing determined approximately by the nested-solids model. Propositions 5–7 summarize Kepler's account of the motion of a single planet. Among other things, proposition 5 includes Kepler's distance rule: "*the true daily arcs of a single eccentric orbit* corresponding with them *have a proportion to each other which is the inverse of the proportion of the two distances from the Sun*" (*HM*, 408; *KGW* VI, 300; emphasis in original). Proposition 6 concerns apparent angular motion as seen from the sun: "*the proportion of the apparent daily arcs on a given eccentric is fairly precisely the square of the inverse proportion of their distances from the Sun*" (*HM*, 409; *KGW* VI, 300; emphasis in original). This proposition is quite important for Kepler's harmonic theory, for he came to hold that planetary harmonies are carried by the apparent diurnal arcs as seen from the sun. Proposition 7 was the basis of yet another archetypal argument against the Ptolemaic system. Because retrograde motion requires considering an infinite number of ratios, such motion is an unsuitable basis for harmonic speculation (*HM*, 410; *KGW* VI, 301).[7]

Propositions 8–13 concern the motions of more than one planet, and thus are most important for calculating the harmonic proportions of planetary motions. Kepler's harmonic law first appears in Proposition 8:

> [I]t is absolutely certain and exact that *the proportion between the periodic times of any two planets is precisely the sesquialterate proportion of their mean distances, that is, of the actual spheres*, though with this in mind, that *the arithmetic mean between the two diameters of the elliptical orbit is a little less than the longest diameter*. (*HM*, 411–12; *KGW* 302; emphasis in original)

Stephenson (1994, 140–41) notes that the qualification following the statement of the third law reveals Kepler's uncertainty as to whether the harmonic law applies to planets with a large eccentricity. Nonetheless, Kepler was pleased, and rightly so:

> For when the true distances between the spheres were found, through the observations of Brahe, by continuous toil for a very long time, at last, at last, the genuine proportion of the periodic times to the proportion of the spheres . . . and if you want the exact moment in time, it was conceived mentally on the 8th March in this year one thousand six hundred and eighteen, but submitted to calculation in an unlucky way, and therefore rejected as false, and finally returning on the 15th of May and adopting a new line of attack, stormed the darkness of my mind. So strong was the support from the combination of my labor of seventeen years on the observations of Brahe and the present study, which conspired together, that at first I believed I was dreaming, and assuming my conclusion among my basic premises. (*HM*, 411; *KGW* VI, 302)

Kepler's harmonic law is shrouded in mystery. The *Harmonice* provides no account of its discovery or justification. Cohen (1960, 138–40) speculates that Kepler simply squared and cubed the numbers until they fit. Gingerich proposes that because Kepler did not discover the harmonic law when searching for a relation between the distances and the periods in the *Mysterium*, he was being not a number juggler but "guided, or rather misguided, by theoretical [physical] considerations" (Gingerich 1975a, 596). Wilson (1978, 238) proposes that Kepler derived the third law from harmonic speculations. Certainly Kepler exploited the third law in his harmonic speculations, and his final harmonic assignments imply it, insofar as they determine the distances and the periods. But these harmonic speculations and assignments cannot be taken as Kepler's only means of either discovering or justifying the third law. As will become obvious, his harmonies were constructed to further justify an already known planetary system, and so could not be taken to determine unknown features of that system. If Stephenson is correct, moreover, Kepler did not yet know whether the third law was approximate or exact. Were it merely a useful approximation, akin in this way to the vicarious hypothesis, it could not be archetypal.

Proposition 9 tells us how to determine true diurnal (linear) motions: "*the true daily arc of each planet must be multiplied by the semidiameter of its orbit*" (*HM*, 412; *KGW* VI, 302; emphasis in original). In proposition 10 Kepler determined the ratio of apparent diurnal arcs derivable by dividing the ratios of the true diurnal motions by the ratio of the distances. Proposition 11 allows one to determine the apsidal distance from mean distances and mean apparent motions. The mean apparent motion is to the apsidal distance as the root of the product of the mean and extreme apparent motions is to the mean distance. Kepler did not use proposition 11 in chapter 9 to determine the apsidal distances, but simply added or subtracted the eccentricity from the mean distances (table 6.9).

Proposition 12 shows how to determine the motions from the ratio of the apparent aphelial to apparent perihelial motion. The mean motion is the geometric mean minus half the difference between arithmetic and geometric means, where the arithmetic mean is the averaged ratio of the apparent extreme motions, and the geometric mean is the square root of the product of apparent extreme motions (*HM*, 413–14; *KGW* VI, 303–4). Later, for example, Kepler harmonically determined the ratio of Saturn's extreme apparent motions to be 64:81. As the geometric mean is 72, and the arithmetic mean 72.50, the mean motion is:

$$72 - \frac{72.50 - 72}{2} = 71.75.$$

In proposition 13 Kepler argued that the ratio of two apparent converging extreme motions (the ratio of the superior planet's perihelial motion to an inferior planet's aphelial motion) is just less than the ratio of corresponding distances to the two-thirds power. Kepler's proof is far from obvious, and because the proposition does not figure in this synopsis, I refer the reader to Stephenson (1994, 143–44). Propositions 6, 8, 10, and 12 show up in Kepler's harmonic derivation of the distances and eccentricities, making them particularly germane to the present discussion.

Kepler did not discuss the physical causes of planetary motion in this section, and mentioned only briefly that the sun is its source, because he considered necessary for assigning harmonies not the physical causes but the spatial or geometrical configurations resulting from them.

In chapter 4 Kepler considered five aspects of planetary motion that might express harmonic consonances: distances from the sun, periodic times, diurnal eccentric arcs, diurnal delays in those arcs, and angles of the diurnal arcs as seen from the sun. He settled on the last, arguing first that diurnal delays are "lengths without motion . . . [and] are not appropriate to be examined for harmonies, because the harmonies are more intimately connected with motion, on account of its swiftness and slowness" (*HM*,

421; *KGW* VI, 310). Second, unperceivable consonances are of little use, and so distances from the sun, periodic times, and diurnal eccentric arcs are likewise unfit to express them. Thus they are expressed in apparent diurnal arcs perceivable from the sun (*HM*, 423–24; *KGW* VI, 311–12).

Kepler's table of consonances is generated by apparent angular diurnal motion at aphelial and perihelial points (table 6.2). Column 1 is the apparent diurnal motion calculated from astronomical theory; column 2 displays the harmonic ratios closest to the extreme motions of two planets; and column 3 is the harmonic ratio closest to the ratio of a single planet's extreme motions. Also included are consonances corresponding to the ratios in column 2.

Kepler uncovered two types of harmonies, those between an individual planet's aphelial and perihelial motions, and those between adjacent planets' diverging, converging, and *eiusdem plagae* motions. Diverging motion is the ratio of the aphelial motion of the superior planet to the perihelial of the inferior. Converging motion is the ratio of the perihelial of the superior to the aphelial of the inferior. *Eiusdem plagae* motion is the ratio of either the aphelial to aphelial or perihelial to perihelial. The ratios were then fitted, in some cases again by brute force, to the nearest harmonic consonances.

Kepler justified the imperfect fit between astronomical and harmonic theories by pointing out that the largest harmonic discrepancies occur between Jupiter and Mars, where the polyhedral model is most accurate (*HM*, 429; *KGW* VI, 316). The point, I take it, is that even large harmonic discrepancies are acceptable so long as the world is still conforming to an archetype. He discounted errors involving Venus and Mercury on the grounds that little was known about their motions. As well, most of the discrepancies would not be discernible to the human ear (*HM*, 427; *KGW* VI, 315).[8]

In chapters 5–8 Kepler considered the consonances as joined, first sequentially in scales, then together in polyphonic harmony. Again, he disregarded errors less than a semitone, promising to reveal their causes later (*HM*, 432; *KGW* VI, 317). He emphasized that the harmonies are found in the aphelial and perihelial movements rather than in those in between. The scales ascribed to the planets in chapter 6 were constructed by filling in the notes between the extreme movements, although the movements between the apsides provide a continuum of tunings rather than the discrete series of notes shown in figure 6.2 (*HM*, 439; *KGW* VI, 322; figure 6.2 is from *HM*, 439). The adjustments of planetary ratios made in chapter 9 disrupt all but Mercury's scale. (Columns 2 and 3 in table 6.4 compare the original harmonies in table 6.2 to the final harmonies assigned in chapter 9).

TABLE 6.2
Harmonies of Apparent Diurnal Motion

1 *Apparent Diurnal Motion*	2 *Harmonies between Two Planets*	3 *Harmonies between the Motion of Single Planets*
Saturn		
Aphelion: 1′46″ a.		a:b = 4:5, major third
Perihelion: 2′15″ b.		
	a:d = **1:3**, b:c = 1:2	
Jupiter		
Aphelion: 4′30″ c.		c:d = 5:6, minor third
Perihelion: 5′30″ d.		
	c:f = **1:8**, d:e = **5:24**	
Mars		
Aphelion: 26′14″ e.		e:f = 2:3, diapente (fifth)
Perihelion: 38′1″ f.		
	e:h = **5:12**, f:g = **2:3**	
Earth		
Aphelion: 57′3″ g.		g:h = 15:16, semitone
Perihelion: 61′18″ h.		
	g:k = **3:5**, h:i = **5:8**	
Venus		
Aphelion: 94′50″ i.		i:k = 24:25, diesis
Perihelion: 97′37″ k.		
	i:m = **1:4**, k:l = **3:5**	
Mercury		
Aphelion: 164′0″ l.		l:m = 5:12,
Perihelion: 384′0″ m.		diapason (octave) and minor third
1:3 = twelfth	1:2 = octave	1:8 = triple octave
5:24 = double octave and minor third	5:12 = octave and minor third	2:3 = fifth
5:8 = minor sixth	1:4 = double octave	3:5 = Major Sixth

Note: Lowercase letters indicate the harmonic ratios closest to the extreme motions of the two planets; boldface indicates harmonic ratios that do not correspond exactly to the ratios of the diurnal motions.

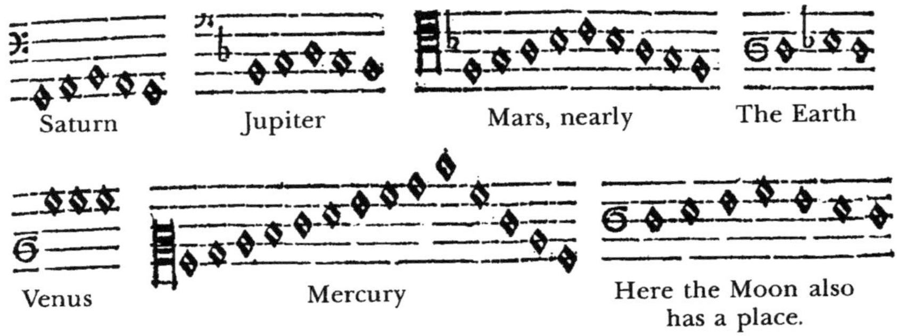

Figure 6.2. Planetary Scales

In chapter 7 Kepler argued that, even though one sextuple harmony found among the planetary movements would be sufficient to indicate that God created the world, he could construct four "skeletal outlines," two *durus* and two *mollis* scales.[9] By skeletal outline Kepler meant a range of notes within which each planet can sound to participate in the chord (*HM*, 443–46; *KGW* VI, 324–27).

It might seem that Kepler's ability to construct not just one but four planetary chords would be a striking victory. But this is not so. Some of the notes were taken from the intermediate motions of the planets, which Kepler believed to lack, or have at most impoverished, harmonic qualities. He held that the best harmonies attach to extreme motions (at aphelion and perihelion). It is important that the best harmonies be possible because Kepler speculated that "if there could occur one single sixfold harmony, or one outstanding one among several, that undoubtedly could be taken as characterizing the Creation" (*HM*, 443; *KGW* VI, 324). Kepler hedged his bets, allowing that one sixfold harmony, and not necessarily the best, could mark Creation. But any harmony less than the best would have been unsatisfactory. Surely God would mark Creation with the best of all possible harmonies. With four acceptable harmonies, none a clear superior, it is not obvious which God would have chosen to start Creation. This is a problem, for the selection seems arbitrary, and Kepler's God never does anything without a reason.

The problem is that Venus's limited range prevents the aphelial motion of Mars and the perihelial motion of Jupiter from participating in the harmonies. In response, Kepler explored the scales possible "if we command this antagonistic lady, Venus, to be silent, that is, if we consider what harmonies there can be, not of all the planets, but at least of the remaining five" (*HM*, 446; *KGW* VI, 327). He was able to construct a skeleton including Jupiter's perihelial motion, but not Mars's aphelial motion. He identified Venus's husband, Earth, as the problem, and constructed another skeleton without Earth and Venus, thus including the aphelial motion of Mars. While Kepler was able to construct harmonies in which all of planets participate, he was not able to accommodate the best harmonies of all the planets at once. One would expect this failure to bother him, because these harmonies do not provide striking evidence that he has succeeded in constructing an archetypal account. But no bother is evident in his closing paragraph in the chapter:

> Therefore, the motions of the heavens are nothing but a kind of perennial harmony (in thought not in sound) through dissonant tunings, like certain syncopations or cadences (by which men imitate those natural dissonances), and tending toward definite and prescribed resolution, individual to the six terms (as with vocal parts) and marking and distinguishing by those notes the immensity of time. Thus it is no longer surprising that Man, aping his Creator, has at last found a method of singing in harmony which was unknown to the ancients, so that he might play, that is to say, the perpetuity of the whole of cosmic time in some brief fraction of an hour, by the artificial concert of several voices, and taste up to a point the satisfaction of God his Maker in His works by a most delightful sense of pleasure felt in this imitator of God, Music. (*HM*, 446; KGW VI, 328)

Chapter 9, perhaps the most significant chapter of the *Harmonice*, is also the most difficult. Kepler fittingly opened it with a prayer to aid understanding. His objective was to show that variation in planetary eccentricities is the result of a rational divine plan. But he discovered that the universe was far less simple than he had believed it to be when he wrote the *Mysterium*. Harmonic archetypes that account for planetary motion turn out to be quite complex, so much so that Kepler described the various harmonic considerations as at war with each other. His recourse was to explain dissonant proportions as the result of more important prior consonances.

Kepler justified the internal tensions of his harmonic construction in forty-nine statements, divided into the *rationes priores* (the first seventeen) and the *posteriores rationes* (the last thirty-two). The prior reasons assign ratios to the planets, Venus and Earth excepted. The posterior reasons struggle with the difficult couple, forcing certain modifications of the ratios assigned by the prior reasons. Kepler wanted to show how the tensions between competing harmonic demands make it impossible to meet them all. Certain adjustments to the harmonic ratios bring them closer to the observed planetary ratios, and so the discrepancy between observed and harmonic ratios is explainable by the conflict between harmonic demands.

The forty-nine statements are further divided into axioms and propositions. The axioms are philosophical principles to which the assignments of ratios must conform, and those of particular significance emphasize either the aesthetic importance of variety, or polyhedral constraints on possible harmonies. The propositions justify both the assignment of consonances and the deviations from them. The assignments are based on the results of chapter 2, where the polyhedra were connected with consonances. The discrepancies are explained by appeal to more important archetypal considerations. If pairs of consonances conflict, the superior is to be left unblemished.

The war between consonances is required to explain discrepancies between apparent diurnal motions and the harmonies closest to them, and to show why some harmonious notes could not be sounded at all. That Kepler discussed the war at length indicates that he still believed the material world represents the archetypal precisely. Had Kepler abandoned this belief, the results in Chapter 4 would have sufficed to show the harmonic nature of planetary motion.[10]

Kepler's first move was to assign polyhedra-based harmonies to converging and diverging motions, and to the private ratios[11] (the apparent aphelial to apparent perihelial motion of single planets). He did not assign all of the ratios in this manner, but calculated those remaining from the following formula:

1. diverging ÷ converging = the product of the private ratios for two adjacent planets.[12]

Most of Kepler's initial assignments correspond to the original harmonies, as shown in table 6.3, which juxtaposes the original harmonies from chapter 4 with the initial assignments from chapter 9. Most of the assignments based on prior reasons matched the original harmonies. The original assignments of a private ratio to Earth, Venus, and Mercury were based on posterior reasons, where Kepler focused on the cause of discrepancies between the harmonies and planetary motion.

As we can see, there is substantial agreement between the harmonies drawn from the polyhedra, and those from the harmonies closest to the observed ratios. If we ignore that the harmonies were linked unconvincingly in chapter 2, this is a promising start for Kepler's archetypal justification of the extreme motions. To see how Kepler justified the agreements and discrepancies together, it will help to trace his assignment of initial harmonies and adjusted harmonies to a particular planet (Mars). Five assignments determine the apparent diurnal motions of Mars: the converging and diverging between Jupiter and Mars, the converging and diverging between Mars and Earth, and Mars's private ratio.

The original assignment of harmonic ratios to converging and diverging motions between Jupiter and Mars was made in proposition 13. Earlier, in assigning ratios to polyhedra, Kepler argued that the tetrahedral ratios should be the squares of the cubic ratios. But because the square of 1:3 (1:9) is not harmonic, and the square of 1:2 (1:4) had already been assigned

TABLE 6.3
Original Harmonies and Initial Assignments

	Original Harmonies			*Initial Assignments*		
	div	*con*	*priv*	*div*	*con*	*priv*
Saturn			4:5			4:5 (11)
	1:3	1:2		**1:3** (8)	**1:2** (8)	
Jupiter			5:6			5:6 (11)
	1:8	5:24		**1:8** (13)	5:24 (13)	
Mars			2:3			18:25 (14)
	5:12	2:3		5:12 (17)	2:3 (15)	
Earth			15:16			14:15 (28)
	3:5	5:8		N/A	**5:8** (15)	
Venus			24:25			35:36 (28)
	1:4	3:5		1:4 (12)	**3:5** (15)	
Mercury			5:12			N/A

Note: Discrepancies are underlined, initial assignments consistent with Kepler's pairing of ratios with the polyhedra in chapter 2 are boldfaced, and numbers of the propositions giving the assignment are bracketed.

to the octahedron, Kepler assigned the next closest harmonic ratios (1:8 and 5:24) respectively to the diverging and converging motions.

In proposition 14 Kepler determined the private ratio of Mars to be 18:25, having determined in proposition 11 that the private ratio of Jupiter is 5:6.[13]

$$\frac{Ja}{Mp} \times \frac{Ma}{Jp} \times \frac{Jp}{Ja} = \frac{Ma}{Mp} = \frac{1}{8} \times \frac{24}{5} \times \frac{6}{5} = \frac{18}{25}$$

(where a and p refer to aphelial and perihelial motion respectively, and J and M to Jupiter and Mars respectively). Mars's original private ratio was 2:3, but this was inconsistent with the ratio derived from Jupiter's. Although Kepler resolved the conflict in favor of the derived ratio, it was already clear that the original harmonies could not all be fit together.

Proposition 15 assigns 2:3 to the converging motion of Mars and Earth, because the proportion of the inner and outer spheres of the dodecahedron equals that of the icosahedron, and so the ratios of spheres for Mars and Earth should be close to that for Earth and Venus. Given 5:8 for the converging motion of Earth and Venus, the nearest harmony that has not already been eliminated is 2:3.

Proposition 17 (the last of the prior reasons) determines that the diverging motion of Mars and Earth is not less than 5:12. To determine the ratio precisely from the formula, Kepler needed the Earth's private ratio, which he had not yet assigned. However, he argued that the diverging motion ought not be lower than 12:25, because Mars's private ratio combined with the converging motion of Mars and Earth is 12:25, and 5:12 is the next greater harmonic ratio.

The five ratios for Mars are adjusted in propositions 40, 41, and 42. In proposition 40, Kepler multiplied the Platonic Limma (243:256) by the ratio of Jupiter's aphelion to Mars's perihelion, because without it the ratio of Saturn's aphelion to Earth's aphelion would be just shy of the five octaves that Kepler determined it to be in proposition 31. The result is 243:2048. This calculation requires readjustments in the related values. In proposition 41, Kepler readjusted the private ratio of Mars accordingly. The diverging motion of Jupiter and Mars was derived in proposition 40, the converging from 13, and Jupiter's private ratio from 38. The resultant private ratio for Mars is 25:36:

$$\frac{Ja}{Mp} \times \frac{Jp}{Ja} \times \frac{Ma}{Jp} = \frac{Ma}{Mp} = \frac{243}{2048} \times \frac{8000}{6561} \times \frac{24}{5} = \frac{25}{36}\,.$$

In proposition 42 Kepler similarly adjusted the diverging motions of Mars and Earth to 54:125.

As we can see, Kepler's assignment of harmonic ratios was terribly complex and, at times, uncompelling. The adjustments do, however, bring the harmonic ratios closer to the observed ratios. Table 6.4 shows the observed ratios and initial harmonies (from Tables 6.2 and 6.3), as well as the final assignments, converted into decimals for comparative purposes (the final assignments will be shown again in Table 6.5 as the ratios Kepler gave). The competing demands of harmony require adjustments to the original harmonies, which requires that some planetary motions be less than perfectly harmonious. This argument to the harmonic archetype is impressive, if complex.

But Kepler wanted to show more than simple correspondence between the archetypal and the observed. He wanted to show that planetary motions could be derived from harmonic ratios. Proposition 48 contains a long and difficult argument to this effect. Here is a sketch. Kepler first obtained the mean distances between the planets and the sun, then the eccentricities of the orbits from the harmonic ratios (table 6.5). These eccentricities and distances, however, are not on a common scale. Kepler then developed a common scale for the harmonically derived private ratios (which express the extreme apparent motions, shown in table 6.6). Using

TABLE 6.4
Observed and Adjusted Harmonic Ratios

	1 *Observed Ratios*			2 *Initial Harmonies*			3 *Final Assignments*		
	div	*con*	*priv*	*div*	*con*	*priv*	*div*	*con*	*priv*
Saturn			.7852			.8			.7901
	.3212	.5		.3333	.5		.324	.5	
Jupiter			.8182			.8333			.8201
	.1184	.2097		.125	.2048		.1187	.2083	
Mars			.6901			.6666			.6944
	.4279	.6663		.4167	.6667		.432	.6666	
Earth			.9307			.9375			.9331
	.5844	.6464		.6	.625		.5832	.6430	
Venus			.9715			.96			.9720
	.247	.5952		.25	.5		.243	.6	
Mercury			.4271			.4167			.4167

this common scale for the extreme apparent motions, he derived the mean apparent motions, both on a private scale (table 6.7), and on a common scale (table 6.8). From the mean apparent motions on a common scale, he derived, also on a common scale, the mean distances. Using the ratios of the mean distances on a common scale to those derived directly from the private ratios, he could convert the eccentricities to a common scale (table 6.9). From the eccentricities and mean distances, he calculated the aphelial and perihelial distances, because the eccentricity measures the amount the extreme distances deviate from the mean. He could then compare the harmonically derived extreme distances to the astronomically derived extreme distances given in chapter 4.

Table 6.5 shows Kepler's derivation of eccentricities from private ratios (*HM*, 484; *KGW* VI, 357). Column 1 cites the propositions yielding the private ratios in column 2. Column 3 converts apparent motions into distances by means of proposition 6 from chapter 3; the apparent diurnal arcs are inversely proportional to the squares of the distances, so one simply finds the square root of the numbers in column 2 to get those of column 3. Kepler gave multiples of the square roots, presumably to avoid decimals in the table. For example, without a multiple of the square roots 8 and 9 for Saturn, half of the difference which yields the eccentricity would be .5.[14] Column 4 is the mean distance from the sun, which is obtained by

TABLE 6.5
Eccentricities from Harmonies

1	2	3	4	5	6
	Private Ratio	*Square Roots*	*Mean Distances*	*Eccentricities*	*Eccentricities If Radius Is 100,000*
Saturn	64	80	85	5	5,882
(by 38)	81	90			
Jupiter	6,561	81,000	85,222	4,222	4,954
(by 38)	8,000	89,444			
Mars	25	50	55	5	9,091
(by 41)	36	60			
Earth	2,916	93,531	95,178	1,647	1,730
(by 28)	3,125	96,825			
Venus	243	9,859	99,295	705	710
(by 28)	250	10,000			
Mercury	5	63,250	80,625	17,375	21,551
(by 45)	12	98,000			

averaging the ratio given in column 3. Column 5 takes the difference between mean and extreme distances, yielding eccentricities, which in column 6 are adjusted to a mean distance of 100,000.

Kepler had astronomically determined the Earth's eccentricity to be 1,800. The harmonically derived 1,730 was simply as close as he could get to 1,800. Jeremiah Horrocks (1618–41) later corrected the astronomical result from 1,800 to 1,730. Ironically, then, though Kepler used 1,800 for the *Rudolphine Tables* (*KGW* X), the harmonically derived eccentricity was more accurate (Wilson 1989, 237, 251).[15] The use of 1,800 instead of 1,730 echoes Kepler's notion, expressed in the *Astronomia*, that while astronomy must be grounded in higher disciplines like physics and metaphysics, it is more closely tied to observation and hence more certain.

Kepler generated distances using harmonic proportions and the harmonic law. His first step was to put the extreme motions expressed by the private ratios on a common scale, although little is said about the significance of this scale.[16] Stephenson (1994, 222–23) argues that this move was intended to show that the extreme motions derived from the private ratios are harmonically intertwined. The scaled numbers were needed to exhibit

TABLE 6.6
Apparent Extreme Motions from Harmonies

1				2	3	4
Harmonic Pairs				*Private Ratios*	*Extreme Motions*	*Prime Factors*
1			Saturn aphelial	64	139,968	$2^6 \times 3^7$
	1		Saturn perihelial	81	177,147	3^{11}
	2		Jupiter aphelial	6,561	354,294	2×3^{11}
	5		Jupiter perihelial	8,000	432,000	$2^7 \times 3^3 \times 5^3$
	24		Mars aphelial	25	2,073,600	$2^{10} \times 3^4 \times 5^2$
	2		Mars perihelial	36	2,985,984	$2^{12} \times 3^6$
32	3		Earth aphelial	2,916	4,478,976	$2^{11} \times 3^7$
		5	Earth perihelial	3,125	4,800,000	$2^9 \times 3 \times 5^5$
	5		Venus aphelial	243	7,464,960	$2^{11} \times 3^6 \times 5$
1	3	8	Venus perihelial	250	7,680,000	$2^{12} \times 3 \times 5^4$
	5		Mercury aphelial	5	12,800,000	$2^{12} \times 5^5$
4			Mercury perihelial	12	30,720,000	$2^{14} \times 3 \times 5^4$

links between the various harmonies expressed by planetary motion. A planet's extreme motion must be proportional to its private ratios and to its harmonic ratio—for example, the ratio of Saturn's aphelion to the Earth's aphelion must be that of 1:32. Together, these constraints require adjustment and readjustment of the ratios between aphelion and perihelion. The extreme motions that Kepler derived are the smallest numbers that satisfy both constraints. Table 6.6 represents this derivation (*HM*, 485; *KGW* VI, 357). Column 1 displays the harmonic ratios between planets, column 2 the private ratios, and column 3 the smallest numbers that fit the above constraints. Column 4 shows the prime factors of column 3.[17]

Once Kepler had the extreme apparent motions, he determined the mean apparent motions. By the third law, he used the mean apparent motions to generate the ratios of the mean distances. He used proposition 12 of chapter 3 to determine mean apparent motions, which "is less than the geometric mean by the same amount as the geometric mean is less than the mean between the two [the arithmetic mean]" (*HM*, 413; *KGW* VI, 303). The arithmetic mean is simply the average of the private ratio. The geometric mean is the square root of the product of the private ratio. Table 6.7 shows Kepler's derivation of the mean motions.[18] Column 1 is the private ratio from table 6.6. Column 4 takes half the difference between the arith-

TABLE 6.7
Derivation of Mean Apparent Motions

1 *Private Ratios*	2 *Arithmetic Means*	3 *Geometrical Means*	4 *Halves*	5 *Mean Motions*
Saturn				
64	72.50	72.00	.25	71.75
81				
Jupiter				
6,561	7,280.5	7,244.9	17.8	7,227.1
8,000				
Mars				
25	30.50	30.00	.25	29.75
36				
Earth				
2,916	3,020.500	3,018.692	.904	3,017.788
3,125				
Venus				
243	246.500	246.475	.0125	246.4625
250				
Mercury				
5	8.500	7.746	.377	7.369
12				

metic and geometric means from columns 2 and 3 respectively. By proposition 12 from chapter 3 one can derive the mean motions in column 5 by subtracting column 4 from column 3.

Because the numbers in column 5 are from the private ratios, they need to be converted to a common measure. The scale used is the proportion between columns 2 and 3 in table 6.6. The numbers derived are those displayed in column 1 of table 6.8.[19] For example, Saturn's private ratio 64:81 is multiplied by 2,187, yielding 139,968:177,147 (columns 2 and 3 from table 6.6). To get Saturn's mean motion on a common scale, then, one multiplies 71.75 by 2,187, yielding 156,917. The same procedure for the other planets will give the remaining numbers for column 1 in table 6.8.

Converting the mean apparent motions into distances requires the harmonic law, which tells us that the cubes of the mean distances are proportional to the squares of the periodic times. One of its implications, by prop-

TABLE 6.8
Mean Motions into Distances

	1 *Mean Motions on a Common Scale*	2 *Inverse Sought among the Cubes*	3 *Mean Distances Sought among the Squares*
Saturn	156,917	29,539,960	9,556
Jupiter	390,263	11,877,400	5,206
Mars	2,467,584	1,878,483	1,523
Earth	4,635,322	1,000,000	1,000
Venus	7,571,328	612,229	721
Mercury	18,864,680[a]	245,714	392

[a] As Stephenson (1994, 224) notes, Mercury should be 18,864,640.

ositions 9 and 10 in chapter 3, is that the cube of the mean distances is inversely proportional to the squares of the mean apparent motions.[20] The inverse mean motions are obtained by standardizing the Earth's inverse motion at 1,000,000, and deriving the others as a function of this inverse and the mean motions on a common scale. For example, the inverse for Mars is (4,635,322 ÷ 2,467,584) × 1,000,000 = 1,878,486 (Earth's motion divided by Mars's and multiplied by Earth's inverse, 1,000,000).[21] For column 3 of table 6.8 Kepler used Clavius's table of squares and cubes in *Practical Geometry*. For example, the mean distance for Mars found among the squares is the cube root of 1,878,486 (123.39), squared (15224.35), and divided by 10, so that the Earth's mean distance will be 1,000.

Having already derived the mean distances and eccentricities from private ratios, Kepler then used the mean distances on a common scale to put the eccentricities on a common scale. (For example, the private mean distance of Saturn is 85, and the common is 9,556. By this ratio, the private eccentricity is 5, the common eccentricity 562. Likewise for the other planets.) From this he derived the aphelial and perihelial distances, where aphelial distance is mean distance plus eccentricity, and perihelial distance is mean distance minus eccentricity. The results are displayed in table 6.9.[22] In column 1 are the private mean distances, and in column 2 the private eccentricities, both from table 6.5. Column 3 gives common eccentricities by the ratio of private mean distances (column 1, table 6.9), and the common mean distances (column 3, table 6.8). Columns 4 and 5 give aphelial and perihelial distances. In brackets are the observed distances from chapter 4.

Stephenson, after comparing the derived aphelial and perihelial distances to the observed distances, comments that Kepler's harmonic deriva-

TABLE 6.9
Distances at Aphelion and Perihelion

	1 *Private* *Mean Distances*	2 *Private* *Eccentricities*	3 *Common* *Eccentricities*	4 *Aphelion*	5 *Perihelion*
Saturn	85	5	562	10,118 [10,052]	8,994 [8,968]
Jupiter	85,222	4,222	258	5,464 [5,451]	4,948 [4,949]
Mars	55	5	138	1,661 [1,665]	1,384 [1,382]
Earth	95,178	1,647	17	1,017 [1,018]	983 [982]
Venus	99,295	705	5	726 [729]	716 [719]
Mercury	80,625	17,375	85	476 [470]	308 [307]

tion is accurate enough that "it is clear why Kepler [felt] no great need to apologize" (Stephenson 1994, 226). Indeed, if we are willing to accept Kepler's assignment of consonances to both polyhedra and planetary motions, the model they yield is impressively accurate.

The Influence of Kepler's Harmonies on His Cosmology

In his Introduction to book V, Kepler commented that the harmonies he discovered were not what he had originally expected, "but in a totally different way, and also at the same time a quite outstanding and perfect way" (*HM*, 389–90; KGW VI, *289*). These changes had a dramatic influence on his cosmology and its philosophical underpinnings. We see variety rather than simplicity in the archetypes, as well as different conceptions of the nature of matter and the perfection of geometry.

Variety and Simplicity

In Kepler's view of the new astronomy, only a complex archetypal system could provide adequate final and formal explanations of planetary motion.

Music theory both allows for a variety of possible relationships and limits the scope of those acceptable. Only some ratios are concordant, only some progressions melodic, and only some chords harmonic. Yet within this structure, a variety of scales and chords are possible. This variety comes at a price, however. Competing harmonies sometimes result in discord, which is why Kepler emphasized that this world is not absolutely perfect but merely the "best possible."

Notice that Kepler was still able to retain a kind of simplicity, the resolvability of complex phenomena into a pattern. Consider, for example, chapter 39 of the *Astronomia*, where Kepler argued that the motions implied by the Ptolemaic model were far too complex to be physically possible. Once the solid spheres were ruled out, the causal mechanism moving the planets in unequal loops would have to be incredibly complex. In the *Harmonice*, Kepler retooled this argument, pointing out that the ratios implied by the Ptolemaic model were inappropriate for harmonic analysis, for they could not be resolved into a pattern (*HM*, 410; *KGW* VI, 301).

Mathematical and Physical Models

Kepler's difficulties in fitting consonances to planetary motions underscores his strong need to complete the archetypal project. His motivations, I have argued, were both methodological and epistemological. The *Mysterium* and the *Astronomia* had left certain tasks unfinished. Still strongly influenced by the scholastic tradition, he needed to provide final and formal explanations of the structure of the universe for his work in the *Astronomia* to count as a proper part of natural philosophy.

He also needed to justify the mathematical measure of libration from chapter 57 of the *Astronomia*. As discussed earlier, Kepler agreed with Aristotle that mathematics alone could not reveal important information about physical change. Mathematics could reveal such information, however, if attached to a qualitative physical or archetypal theory. The magnetic analogy on which the measure of libration was based was inadequate. But the sum of sines model was sufficiently aesthetically pleasing to be a candidate for archetypal justification. It unified the nonuniform motion of planetary libration. In the *Astronomia*, the elegance of the sum of sines model was an indication of its candidacy for measuring a force. In the *Harmonice* it turns into an extremely important by-product of the archetypes. Although harmonic archetypes determine only the extreme motions, Kepler's God would have created an orderly means by which the path between extreme motions could be traversed. In addition, because the eccentricity of each planetary orbit is different, a unifying measure that produces various harmonies from different starting points would be especially aesthetically pleasing. If the mean distances, aphelial motions, and perihelial motions are determined archetypally, then on the assumption that all other motion is methodically generated, all motion is archetypally determined.[23] The orderly generation of planetary libration freed Kepler to examine only extreme movements for harmonic consonances:

> And because all these [the arcs, the delays in equal arcs, and the distance of the arcs from the sun or the intervals] are as it happens variable in the case of the planets, there can be no doubt that if they have been assigned any geometrical beauty, by the sure design of the Maker, they acquire it at their extremes, as at their distances in aphelion and perihelion, and not so much at the mean distances in between. For given the proportions of the extreme distances, the design does not need to fit the intermediate proportions to a definite number; *for they follow automatically, by the necessity of the planetary motion from one extreme, through all the intermediate points, to the other extreme.* (HM, 420; KGW VI, 308; emphasis added)

Did this orderly generation increase Kepler's confidence in his account of libratory motion? The "physical" version of his measure of libration (that the strength of the attraction or repulsion is measured by the sine of the coequated anomaly) entailed that the threads that interfere with the sun's motive force, causing the planet to librate in its orbit, cannot be identified with the axial motion of the planet. When writing the *Astronomia*, Kepler felt it unlikely that any other portion of the planetary body was suitable for being identified as the magnetic threads, and suggested that there is a need for planetary minds to guide the planet (*AN*, 559–60; *KGW* III, 355–56). Yet in the Epitome, published after the *Harmonice*, Kepler stated with casual assurance that the threads must be appropriately oriented, and therefore must be under the crust of the planet's surface (*Epitome*, 937; *KGW* VII, 335–36). He no longer seemed concerned with the need to identify the threads with some observable property of the planet. Perhaps he was more confident in his hunches about the magnetic properties of planets.[24] The archetypes did after all further support his mathematical measure of libration. The physical hunches on which his mathematical model was based were indirectly supported by increased trust in that model. This confidence suggests that Kepler believed attaching a mathematical model to an archetype would allow one to draw physical conclusions from that mathematical model.

Geometry and Harmony

In the envoi to chapter 9, Kepler compared harmony and geometry in a manner strikingly unlike the Kepler so far encountered, so much so that I will quote the passage at length:

It was right that in the genesis of the distances the solid figures should give way to the harmonic arguments. . . . *For where there is a choice between different things which do not allow each other to have sole possession, in that case the higher are to be preferred, and the lower must give way, as far as is necessary*. . . . *But harmonic decoration is as far above the simple geometrical as life is above the body, or form above matter*. . . . *But the five solid figures, in virtue of the word itself, relate to the spaces of the regions, and to the number of them and of the bodies; but the harmonies to the motions. Again, as matter is diffuse and unlimited in itself, but form is limited, unified, and itself the boundary of matter; so also the number of geometrical proportions is infinite, the harmonies are few. For although even among the geometrical proportions there are definite degrees of limitation and shape and restriction, and not more than three can be formed by the ascription of spheres to regular figures, yet even to these is attached an accidental property in common with all the rest, that is the presupposition of an infinite possible division of quantities*. . . . *But the harmonic proportions are all expressible, and the terms of all of them are commensurable. Also, they have been taken from a definite and limited class of plane figures*. . . . *Therefore, as matter strives for form, as a rough stone, of the correct size indeed, strives for the Idea of the human form, so the geometrical proportions in the figures strive for harmonies*. (*HM*, 488–89; *KGW* VI, 360–61; emphasis in original)

Here it seems Kepler was ranking harmony above geometry, comparing the harmonic to form and the geometric to diffuse and unlimited matter. He went on to speak of his polyhedral hypothesis:

Of course in this house of the cosmos I was looking for nothing but the stones—of more elegant form, but of a form appropriate to stones—not knowing that the Architect had shaped them into a fully detailed effigy of a living body. So little by little, especially in these last three years, I came to the harmonies, deserting the solid figures over fine details. (*HM*, 490; *KGW* VI, 361; emphasis in original)

This seems a dramatic shift from the *Mysterium* and even books I to IV of the *Harmonice* where Kepler wrote that quantity is part of God's essence, and that God had no other reason to create physical powers except to instantiate quantity (*MC*, 93–95, 123; *KGW* I, 23–24, 37–38; *HM*, 9, 139, 295, 298–304; *KGW* VI, 9, 100, 215, 218–23). In addition, although Kepler equated quantity with form in the *Mysterium* and books I–IV of the *Harmonice*, he ceased to do so in book V.

So Kepler's astronomical discoveries influenced their metaphysical underpinnings. He needed to explain why the physical world did not conform to its geometrical pattern, since it seemed obvious that matter would conform if able:

The placing of the solid figures among the planetary spheres, if it is unrestricted, and not prevented by the necessities of preceding causes, ought in perfection to follow the analogy of the geometrical inscriptions and circumscriptions, and therefore the terms of the proportion of the inscribed spheres to the circum-

> scribed spheres. *For nothing is more fitting than that the physical inscription exactly represents the geometrical, as a printed work does its type.* (*HM*, 480–481; *KGW* VI, 354; emphasis in original)

The nonuniform elliptical motion of Mars did more than lead Kepler to reject the circle as the basis for planetary models. It also led him eventually to reject pure geometry as the determinant of material form. When writing the *Astronomia*, he had not yet reached this conclusion. Recall rejection of the first physically defined ovoid model of chapter 45 on the grounds that it is not geometrically determinable: "This is not just a fault in our understanding, but is utterly alien to the primeval ordainer of the planetary courses: we have hitherto found no anticipation of such lack of geometry in the rest of his works" (*AN*, 489; *KGW* III, 310). Yet in book V of the *Harmonice*, Kepler writes:

> Hence just as the bodies of animate beings have not been made, and a mass of stone is not usually made, according to the pure norm of some geometrical figure, but something is removed from the external round shape, however elegant (though the correct amount of bulk remains) so that the body can take on the organs necessary to life, and the stone the likeness of an animate being, similarly also the proportion which the solid figures were to prescribe for the planetary spheres, as lower, and having regard only to a body of a particular size and to matter, must have given way to the harmonies, as much as was necessary for the former to be able to stand close and to adorn the motions of the globes. (*HM*, 490; *KGW* VI, 362)

Thus, in Kepler's revised view, an acceptable model need not be geometrically determinable since the harmonic can interfere.

One wonders whether it was easy for Kepler to revise his view of geometry. I suspect not. The distinction between concordant and discordant ratios is explained in book III by grounding concordances in geometry. Kepler had held this position for quite some time. In a letter to von Hohenburg, Kepler argued that geometry is the foundation of the universe, and that corresponding to every geometrical relation there is an empirical relation and a musical consonance (January 1607; *KGW* XV, 392). I suspect that although Kepler felt he needed the somewhat messy harmonic-geometric archetypal model, he preferred purely geometrical archetypes.

The new relationship between geometry and harmony influenced Kepler's conception of the material world. Earlier, in the *Mysterium*, Kepler believed God could create matter to perfectly instantiate the archetypes. Because matter *could* conform to the archetypes, it did conform. God would not create anything otherwise: "For if they are knowable, then they can enter the Mind and into the shaping of the archetype; but if they are unknowable (in the sense which has been explained in Book I) then they have remained outside the Mind of the eternal Craftsman" (*HM*, 139; *KGW* VI, 100–101). But later in the *Harmonice* Kepler spoke as though the material imposes limitations of its own. Matter could not conform to a pure geometrical archetypal model if it was to "take on the organs necessary to life" (*HM*, 490; *KGW* VI, 362). This statement has two possible implications, one ontological, the other epistemological. First, limitations imposed by the nature of matter imply that the divine blueprint is untidy, suggesting that the material determined, to some extent, the nature of the archetypes, whereas in Kepler's earlier work, only the converse held. The second implication is more problematic. It could be the case that matter is sufficiently unruly as to block a complete and precise archetypal account. Now, the failure to conform to an archetype suggests either that the archetype is interfered with by some other archetype, or that matter is refusing to behave itself. In Kepler's earlier thought, because the material perfectly instantiates the formal, archetypes can be empirically tested and used as theoretical support for *both* physical and astronomical theories. Matter's new unruliness makes it fail to instantiate the archetypes perfectly, further complicating the testing process. The deviation of observations from theory could be due to either the incorrectness of the theory, observational error, or matter's unruliness. There is no longer as tight a fit between archetypal theories and data, or between physical and archetypal theories. As a result, Kepler's earlier position, that a properly established theory is more trustworthy than observations (*MC*, ch. 18), is no longer tenable.

If I am right, Kepler would have found his new cosmological vision uncomfortable. The God he was communing with in studying nature turned out to be more limited, and less an aesthetician than he had believed when writing the *Mysterium*. While his exuberant boasts[25] can still be taken as sincerely expressing the belief that he had uncovered the divine blueprint, they cannot be so easily taken as sincerely expressing delight in this new-found universe. While accessing the divine blueprint was no mean feat, the consequences of this knowledge were problematic. Kepler's belief that one could obtain exact knowledge of planetary motion and its causes was undercut, for as I have argued, precise archetypal-physical correspondence was the foundation of this belief, making it difficult to take his ecstatic assessments of the discovered harmonies at face value. Kepler's "mind's favorite child" (Caspar [1959] 1993, 288) was a rebellious one.

7

The *Epitome astronomiae Copernicanae*: Kepler's Mature Physical Astronomy

> It has been ten years since I published my *Commentaries on the Movements of the Planet Mars* [*Astronomia nova*]. As only a few copies of the book were printed, and as it had so to speak hidden the teaching about celestial causes in thickets of calculations and the rest of the astronomical apparatus, and since the more delicate readers were frightened away by the price of the book too; it seemed to my friends that I should be doing right and fulfilling my responsibilities, if I should write an epitome, wherein a summary of both the physical and astronomical teaching concerning the heavens would be set forth in plain and simple speech.
> (*Epitome*, 845; *KGW* VII, 251)[1]

THE *EPITOME ASTRONOMIAE COPERNICANAE*, contains much of the material from Kepler's earlier works, yet stands out in two very important respects. First, it was written for a more general audience, and indeed it gained a relatively wide readership.[2] Second, Kepler's mature physics, metaphysics, and astronomy were presented together for the first time. As a result, it is an invaluable resource for exploring the evolution of Kepler's thought, fleshing out his conception of the relationship between physics, metaphysics, and astronomy, and—since the *Epitome* was intended as a textbook—uncovering what Kepler believed he needed to do to promote his new astronomy.

The *Epitome* comprises seven books.[3] Books I, IV, and V are of particular interest here, for it is in these that methodological issues are addressed directly and the archetypes appear most frequently.[4] In book I, Kepler introduced the discipline of astronomy and the nature of its subject matter. He also defended the thesis that the Earth rotates. Book IV concerns the archetypal and physical structure of the universe, and book V the mathematical representation of that universe.

The *Epitome* provides evidence that Kepler still viewed physics, astronomy, and metaphysics as parts of a comprehensive philosophical system, one that could overcome the conceptual difficulties in developing a physical astronomy. As we have seen, however, Kepler's philosophical system was starting to buckle under the strain of accommodating the complexity of the physical world. Consequently, we see two conflicting trends in the *Epitome*. The physical world conforms to the archetypal, and this assumption is used to support astronomical hypotheses, yet in some cases, owing to its nature, matter rebels against such conformity. My interest is in the evolution of Kepler's thought, and so for contrastive purposes I concentrate primarily on Kepler's treatment of topics addressed earlier in the *Mysterium* and *Astronomia*. I do not, for example, discuss Kepler's lunar theory, which is thoroughly examined by Stephenson (1987, 175–201).

The structure of this chapter, then, mirrors the arc of my argument thus far. The first main theme is Kepler's use of the archetypes, the second how the complexity of the physical world jeopardized Kepler's world view.

In book I, Kepler's goal was to counter the Aristotelian contention that it is in the Earth's nature to remain stationary. Fighting fire with fire, he argued that on the most natural conception, the Earth rotates on its axis in addition to moving around the sun. To defend this, he needed to clear up some misconceptions on the nature of astronomy and on the proper treatment of observations.

Book IV, on the physical and archetypal causes of the universe, is divided into three main sections. In part I the archetypal system is presented as unproblematic and is recapitulated from the *Mysterium* with one notable difference in emphasis. In the *Mysterium*, Kepler's physical intuitions were nascent. His primary goal was to argue for the aesthetic advantage of the Copernican model over the Ptolemaic. Here, Kepler's goal was more complex. In addition to metaphysically justifying the Copernican system, he aimed to present his celestial physics as a natural consequence of his cosmology. That Kepler kept his physics in mind is particularly obvious when he added to the Trinity and polyhedral archetypes an archetypal determination of planetary densities, which he later used in a derivation of the harmonic law. What I find particularly interesting about part I is that, because it is clearly designed to gain converts from a general audience, Kepler believed that his physics would be more readily accepted if it followed naturally from the archetypes.

Parts II and III concern Kepler's celestial physics, including material from the *Astronomia*. There are significant differences between the two presentations, however. In the *Astronomia*, Kepler frequently used the notion of mind as cause to explore the properties of a causal system. In the *Epitome* he ruled out this method by arguing that minds as causes would produce different kinds of effects than material causes (the pattern produced by a mind would be more elegant). This argument is, I suggest, in part a response to difficulties in maintaining a correspondence between the rational and the material. Another of Kepler's means for dealing with this was to suggest that, whereas isolated causes would produce archetypally significant relations, compound causes might not. Thus, the failure of a physical system to live up to its archetype actually provides information about the complexity of the causal structure.

The subject of book V is the derivation of mathematical astronomy from the physics of book IV. My focus here will be Kepler's libration theory. Once again he encountered difficulties in making the mathematical and physical theories consistent. Although his preference was to ground quantitative theories in qualitative ones (a requirement he developed in the *Apologia*), this problem spurred him in a very useful direction. He finally developed an explicit account of how it is that a mathematical model can represent a physical system.

Evident throughout the different discussions of books I, IV, and V is the influence of the archetypes on Kepler's construction and defense of his astronomy. Archetypal reasoning, in various forms, played a systematic series of roles in Kepler's physical and methodological thought, especially in the methodological changes motivated by the difficulties arising from the *Harmonice*.

The Nature of Astronomy and Observation

Because the *Epitome* was intended as an introductory text, it is natural for it to open with general questions about the subject matter, the very first of which is "What is Astronomy?" This seemingly innocuous opening is really quite loaded, as Kepler's answer was atypical for its time. He conceived astronomy to be the study of both the heavens as seen from a moving Earth and the causes of these appearances. Thus it involves recording the appearances as they occur, and constructing models that predict and retrodict them (*KGW* VII, 23). He also argued that astronomy requires linked knowledge from other disciplines, including physics, geography, history of celestial motion, meteorology, optics, mathematics, and the study of instruments.

Having advocated an interdisciplinary approach, Kepler then addressed how the various relevant disciplines relate to each other. He reemphasized the need both to ground astronomy in physics and metaphysics, and to avoid extradisciplinary use of that which pertains to a discipline considered individually (*KGW* VII, 25). This discussion indicates Kepler's deep concern to legitimate his particular mixing of disciplines. As discussed earlier, Kepler took seriously Aristotle's strictures on interdisciplinary work. Notice that some of the disciplines listed, like optics and the construction of astronomical instruments, were already considered obvious necessary auxiliaries to astronomy. I suspect that Kepler's purpose was to suggest subtly that astronomy was never rightly considered an isolated discipline. The right approach, then, is to delineate the relationships between the relevant disciplines, rather than to try fruitlessly to keep astronomy isolated.

After claiming that knowledge of physics and metaphysics is necessary for studying astronomy, Kepler argued that his interdisciplinary approach would alter (correctly) the interpretation of observations. In the dedicatory letter for the first three books of the *Epitome*, Kepler announced his goal to filter out prejudices against "the most important and dignified hypothesis [that the Earth rotates]" (*KGW* VII, 8), prejudices arising from facile interpretation of observations. The inference from the apparent motion of the heavens around the Earth to the conclusion that the Earth is the central stationary body would fail if properly processed by reasoning about physical and metaphysical causes. The inference that there are only two principle parts of the universe (heaven and Earth) likewise arises from the failure to process these observations properly (*KGW* VII, 26–28). Causal reasoning (in the metaphysical sense) reveals that the universe has three principle parts, without which it could not represent the Trinity.

Proper reasoning should lead us to the conclusion that, despite the apparent rotation of the stellar sphere, it is rather the Earth that rotates on its axis. For one thing, although Aristotle claimed the nature of spherical objects is to rotate, the cause of this rotation is not clear. In the case of the stellar sphere, the cause cannot be the motive force from the sun, because the stellar sphere appears to be moving too fast. Moreover, Kepler maintained, there are many reasons to suppose that the body of the Earth is well suited for rotation on its axis. For one thing, its body is much like that of an animal, producing heat, water, minerals, and so on. As an animal's body is well suited to motion, so too then is that of the Earth.

In addition, inspired by William Gilbert, Kepler argued that the Earth has magnetic fibers running parallel to the equator. Together, the direction and magnetic nature of the fibers makes them suitable for producing diurnal rotation. On the animal analogy again, Kepler reasoned that the Earth may well have a soul that uses the magnetic fibers to produce regular diurnal motion. No such analysis of the cause of the rotation of the stellar sphere is available. Thus filtering the appearances through causal reasoning reveals different causes than the appearances themselves suggest (*KGW* VII, bk. I, pt. V).

Kepler went to great lengths to show how his system implies the appropriate observations. Because the observations alone do not decide between a rotating Earth and a rotating stellar sphere, however, Kepler needed some such metaphysical and physical considerations as those just described to make his case. This requirement was not inappropriate. After all, standard arguments against the axial rotation of the Earth appealed to metaphysical and physical premises.

One also sees the use of archetypal reasoning in Kepler's argument that the universe is finite rather than infinite. The basic argument is as follows. If the universe were infinite, then the stars would have to be uniformly distributed, because it is impossible to impose any other geometrical order on infinite space. This distribution entails an appearance, however, that we do not observe. If the stars were evenly distributed, we would observe only a few, evenly spaced out, nearby (Kepler claimed there would be no more than twelve of the closest group of stars in order for them to be both evenly spaced from the Earth as well as from each other). The next group would be twice as distant, the next three times as distant, and so on. Because of the quickly increasing distances of successive groups, only a few of the groups would be visible. Instead, what we see is a great number of stars, and so the distribution of stars cannot be infinite (*KGW* VII, 42–46). Clearly, without the use of archetypal reasoning, Kepler would have been unable to generate predictions from the hypothesis of an infinite universe.[5]

Archetypes as a Prerequisite for Physics

Prior to his "war on Mars," Kepler had an archetypal justification for every feature of his then undeveloped astronomy (i.e., the Copernican order of planets, and the use of the true sun). Kepler's war resulted in nonuniform elliptical orbits which, prior to writing the *Harmonice*, had no archetypal justification. As argued earlier, Kepler believed that a complete, fully supported description of nature would include archetypal explanations, and the *Harmonice* was an attempt to fill in this lacuna vis-à-vis the nonuniform ellipse. In the introduction to book IV of the *Epitome*, Kepler alluded to the need for precisely this sort of justification: "But if, however, even this inquiry [into the harmonies] is accused of being esoteric: I indeed confess that the head of astronomy is struck off" (*Epitome*, 850; *KGW* VII, 255). Kepler went on to say that astronomy could function "without a head" so long as one is only interested in predictions. If one is interested in the true nature of the heavens, then investigation of the archetypes is required.

The preceding quotation makes it clear that Kepler viewed archetypal thinking as critical to a realist physical astronomy. Just what he conceived that critical role to be is not obvious, and so we need to take a closer look at book IV of the *Epitome*. Although Kepler continued to use the archetypes for metaphysical and epistemological purposes, in certain respects their role was markedly different from what it had been in previous works.

Book IV and the Archetypes

The title page of book IV announces an explanation of the archetypal and physical causes of the magnitudes, proportions, and motions of celestial bodies. There are three sections, providing explanations of, in turn, the causes of the configuration of the universe, the primary causes of planetary motion, and the causes of the irregular planetary motions. The archetypes, though present throughout, do not play the same role in all three etiologies.

Part I

Kepler intended part I to win converts to his version of the Copernican system. He tried first to persuade the reader that, despite displacing us from the center of the universe, the Copernican system was created to benefit humanity. He then laid the groundwork for the celestial physics presented in parts II and III.

To accomplish this, Kepler turned the aesthetic disadvantage of the Copernican system into an anthropocentric advantage. The system requires irregular spacing of planetary orbits, and although this may disturb the mind, the disturbance is more than outweighed by consequent benefits: "And so this lay-out of his [Copernicus's] urges the speculative mind to spurn matter and the contiguity of spheres and to look towards the investigation of the formal lay-out or archetype, with reference to which the intervals were made" (*Epitome*, 861–62; *KGW* VII, 265). Despite the prima facie appeal of regular spacing, the irregular spacing spurs the mind to consider higher ordering principles. Thus in this among other respects, the Copernican arrangement is of greater benefit than the Ptolemaic. In a similar vein, Kepler claimed that eclipses were created to inform us of celestial phenomena (*Epitome*, 876; *KGW* VII, 279–80).

In addition to the foregoing arguments, Kepler recapitulated archetypal considerations from the *Mysterium*, in particular the Trinity and polyhedral archetypes. Here we get the clearest sense of how these archetypes are part of Kepler's physics. His purpose was to determine archetypally the topological features of the universe and to set the stage for showing that the archetypes are consonant with, and indeed necessary for, physical reasoning. To put the point another way, the general argument in book IV is that in conjunction with ordinary physical causes the archetypal will produce an empirically adequate model of the world. Part I addresses the world's archetypal structure.

One significant difference between the discussion here and that in the *Mysterium* is that here the physical arguments to come were already clear in Kepler's mind. For example, in his discussion of the Trinity, Kepler stated much more clearly that the sun in representing God was well suited to producing motion. To show that the sun is analogous to God, he compared the sun's faculties with those of the soul. "The perfection of the world consists in light, heat, movement, and the harmony of movements. These are analogous to the faculties of the soul: light, to the sensitive; heat, to the vital and the natural; movement, to the animal; harmony, to the rational" (*Epitome*, 854; *KGW* VII, 259). The argument is that because every perfection of the world is in the sun, and these perfections are analogous to the soul, the sun is the appropriate body to represent God the father in the Trinity. Kepler continued with arguments that as producer of light, heat, movement, and harmony of movement, the sun ought to occupy a central position, such a position being dignified, such powers being evenly distributable, and so on.

Notice that the analogy here is not just between God as primary mover and the sun as primary mover, but also between sun as mover and soul as movers of physical bodies. The concept of God as primary mover is abstract and metaphysical, as far removed as one could want from commonsense physical causes. An animating soul, however, is a far more ordinary kind of cause, one that we seem, if we are dualists, to experience every day. Kepler was working to bring the notion of sun as cause out of the realm of the extraordinary into the realm of ordinary, natural causes.

In section IV of part I, Kepler offered a demonstration of the relative sizes of the Earth, sun, and moon. Although Kepler's purpose was to detail the structure of the world as God created it, his determination also had practical applications (computing eclipses, for instance). This sort of information cannot be computed securely from observations because the possibility of error is too great. Thus Kepler turned to archetypal reasoning. Although the sun may be the site for setting up ratios of the harmonies, the first step in determining the ratios should be on the basis of how things appear from the Earth since the Earth is not only home for the "measuring creature" but is the middle body between the sun and Saturn. Because light is first in the order of creation, and the sun the principle body, the first body to consider is the sun's.

Observations both ancient and modern up to Kepler's time indicated that when the sun is furthest from the Earth, it appears as $^1/_{720}$ of a circle, or ½°, with a radius of ¼°. To ensure the correctness of this measurement, Kepler justified it archetypally. Recall that harmonic intervals can be represented as ratios of the length of a monochord's string. Kepler argued that 720 is the smallest number of parts into which the length can be divided such that every interval in both major and minor modes can be represented by a whole number. (For example, by placing the division at unit 540 of a string divided into 720 units, one would get the interval of a fourth, 4:3.) This reason explains the archetypal nature of the number 720. Kepler would have preferred a geometrical rather than harmonical explanation, but there was no geometrical way to divide a circle into 720 parts (*Epitome*, 874; *KGW* VII, 277).

From this supposition Kepler determined that the greatest distance between Earth and sun d_{se} is just over 229 radii of the sun. That is, where d_{se} is the distance to the sun, and the radius of the sun r_s is ¼°:

$$d_{se} = \frac{r_s}{\tan ¼°} = 229.18166 r_s \, .$$

Because the Earth is the home of the "measuring creature," it should be the standard of measurement. In other words, the ratios of volumes of bodies should be set to its body, the ratios of distances to its radius. It is appropriate for archetypal reasons that the ratio of the Earth's body to the sun's is equal to the ratio of the Earth's radius to its distance from the sun. This equivalence allowed Kepler to derive proportions of the Earth. If the radius of the sun r_s is 100,000, then the distance to the sun d_{se} will be 22,918,166. Because the volumes of the bodies are as the cubes of their radii, and the ratio of the body of the Earth r_e^3 is to the body of the sun r_s^3 as the ratio of the radius of the Earth r_e is to the distance to the sun d_{se}:

$$\frac{r_e^3}{r_s^3} = \frac{r_e}{d_{se}} \, .$$

To get the Earth's radius:

$$\frac{r_s^3}{d_{se}} = r_e^2 \, .$$

Because r_s is 100,000 and d_{se} is 22,918,166, r_e will be approximately 6,606. Because $3{,}469\tfrac{1}{3} r_e = 22{,}918{,}166$, then $3{,}469\tfrac{1}{3} r_e^3 = r_s^3$. Therefore, the body of the sun is approximately 3,469 times larger than that of Earth.

The ratios of bodies and distances, though determined archetypally, had observational consequences. As Kepler pointed out, by setting the distance at approximately 1,200 times the radius of the Earth, the ancients set the parallax to 3′. In reasoning that Mars was closer than the sun and would therefore have a parallax of greater than 3′, Brahe was unable to determine the parallax of Mars at all (Gingerich and Voelkel 1998). Therefore,

Kepler argued, on the archetypally justified assumption that the diameter of Mars is not smaller than that of Earth, Mars is further away than 1,200 r_e. The sun, consequently, is further away still.

The large sun-Earth distance and consequent large sun radius also befit the Trinity analogy. Kepler argued that an implication of the Trinity analogy is that the size of the sun is greater than all the planets put together (*Epitome*, 876; *KGW* VII, 279). He observed that this implication is appropriate because the sun moves all the planets. He did not, at this point, discuss why the sun would need to be larger to move the other bodies, but once again, though he did not offer any physical arguments, he was thinking physically. His point was that an archetypally appropriate universe is also physically appropriate.

Kepler used the same approach in determining the relative size of the moon. The visible diameter of the moon at its greatest distance from the Earth is also $^1/_{720}$ of the circle for two reasons. First, the number itself is special, as already argued. Second, the moon was created just the right size to block out the sun exactly during an eclipse, an indication of God's design to teach us of the universe (figure 7.1).

As with the sun, it is appropriate for archetypal reasons that the ratios between the body of Earth and that of the moon be equal to the ratio between the distance to the moon and the radius of the Earth. Once again, the distance is approximately $229r_m$. So if the radius of the moon is set to 100,000, then d_{me} is 22,918,166. By the above ratios:

$$\frac{r_e^3}{r_m^3} = \frac{d_{me}}{r_e}$$

or:

$$d_{me} = \frac{r_e^4}{r_m^3}.$$

Because r_m is 100,000, and d_{me} is 22,918,166, then r_e will be 389,085. Because just less than $59r_e$ = 22,918,166, slightly less than $59d_{me} = r_e$. This result was observationally significant because Brahe had calculated the greatest distance to the moon to be between 59 and $60r_e$.

Kepler also got an archetypally significant result in that it could be demonstrated that the distance from the Earth to the moon is a mean proportion of the distance from the Earth to the sun and the radius of the Earth:[6]

$$\frac{d_{se}}{d_{me}} = \frac{d_{me}}{r_e}.$$

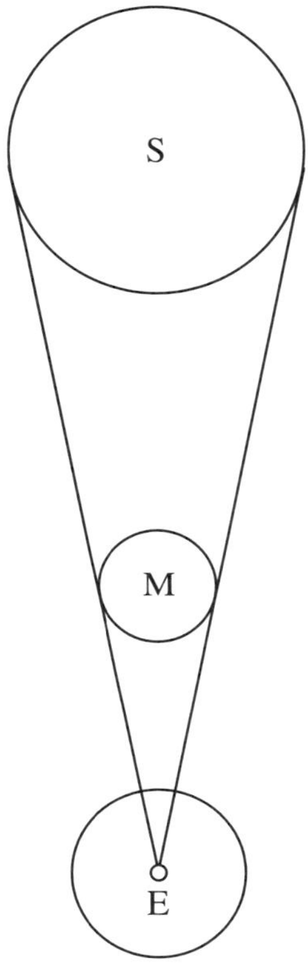

Figure 7.1. Determining the Proportions of the Sun, Moon, and Earth

The determination of the other planetary bodies, however, follows a somewhat different procedure. It is still the case that for archetypal reasons, the order of magnitudes of the bodies follows the order of magnitudes of the radii of the orbits. There are three ways in which this can be construed.

The first is to hold the diameter of the bodies proportional to the distances. So, because Saturn is approximately 10 times further than the Earth from the sun, the diameter of Saturn is approximately 10 times that of the Earth. Thus the surface area of Saturn is approximately 100 times that of the Earth, and the volume of Saturn is approximately 1,000 times that of the Earth. Kepler rejected this possibility immediately on both archetypal and empirical grounds.

The second way of construing the order of magnitudes relation is to hold the surface areas proportional to the distances. If so construed, the surface of Saturn is approximately 10 times that of the Earth, and its volume approximately 30 times. This calculation puts the ratio of volumes at

$^3/_2$ the power of the distances. This is Kepler's preferred construal, presumably because it allows a straightforward derivation of the third law from the archetypal assignment of the ratios. He nonetheless rejected this construal, because observations did not support it.

The third way is to hold the volume of Saturn proportional to the distances, which sets it at approximately 10 times that of the Earth. Kepler then argued that the ratio of $^3/_2$ powers of the distances will be maintained in the body of the planets. Presumably this means that the density of matter contained in the body is responsible for retaining the $^3/_2$ proportions.

That Kepler's archetypal arguments were motivated by physical considerations is most evident in his archetypal determination of planetary densities. He had to determine the relative densities archetypally, for he could determine them neither by observation nor by inference from some physical theory. Yet nothing in his archetypal determination of densities is as aesthetically interesting as the polyhedral model. The principles by which he computed the densities are as follows.

First, by archetypal reasoning, when there are a number of bodies, they must differ in both amount of matter contained and the ratio of matter contained to body size. This reasoning is an extension of the principle of variety introduced in the *Harmonice*. Kepler said little here to justify this requirement, although, as in the *Harmonice*, he admitted that empirical adequacy (in this case, the third law) is a primary reason for holding this requirement.

The second principle is that the closer a body is to the sun, the denser it must be. His argument for this principle verges on the physical: "For the sun itself is the most dense of all bodies in the world, and its immense and manifold force, which could not exist without a proportionate subject, bears witness to this thing" (*Epitome*, 880; *KGW* VII, 283). This argument, though it verges on the physical, is metaphysical in nature. The claim is that there must be a relation of appropriateness between a cause and its subject. Kepler demurred from discussing the physical consequences of the situation being otherwise. For example, he did not comment on how slowly the most distant planet—Saturn—would move if it were extremely dense.

Kepler's third principle is simply that in order to get right the proportions between the bodies and the distances, the density should be fixed in the following way:

> [I]t happens that the ratio of the amount of matter and the ratio of the density are both the ratio of the square roots of the intervals from the sun, and that thus the amount of matter and the density participate equally but inversely in that ratio: hence the amount of matter is greater, hence the density of the same great body is less: and that is the best mean [mediatio] of all. (*Epitome*, 881; *KGW* VII, 284)

TABLE 7.1
Planetary Densities

Saturn	324	Hardest precious stones
Jupiter	438	Loadstone
Mars	810	Iron
Earth	1000	Silver
Venus	1175	Lead
Mercury	1605	Quicksilver

So that we may reserve gold—whose density in this proportion is 1,800 or 1,900 for the sun (*Epitome*, 882; *KGW* VII, 284)

Kepler needed to specify a relationship between distances and bodies that is related to the third law, and he needed a means of doing so that, though not supported by observation, has not been ruled out by observations. But Kepler had another archetypal argument up his sleeve. If we take the densities thus obtained, and find the mean proportional between the mean distances to the sun, then find the least-common number, we get the following numbers. These numbers yield a correspondence between various materials and the planets as displayed in the table Kepler drew up (table 7.1).

Kepler did not explain the significance of this table, although the reservation of gold for the sun is clearly of archetypal significance. This correspondence is not nearly so interesting as his determinations of the ratios of the Earth, sun, and moon, and certainly not as aesthetically pleasing as his polyhedral archetype. Notice the difference between the discussion here and that of the polyhedra. Where Kepler dwelled at length on the aesthetic qualities of the polyhedra, here he quickly moved on as though leaving the scene of an accident. His purpose here was not to delight the reader with some beautiful mystery he thought he had uncovered. Instead, he was setting the stage for showing that the planet's natural inertia (by this Kepler meant resistance to motion) in conjunction with the force of the sun would produce the periods required by the third law. Notice, however, that Kepler was not yet showing how the third law follows from the sun's force in conjunction with distances, lengths of orbits, and densities. Rather, he was still stage setting, arranging the archetypes so that the forthcoming physical arguments would follow as natural consequences.

Kepler moved on to determine the ratios between the three principle parts of the world as set by the Trinity archetype. In book I he argued on a number of grounds that the region of the stars could not be of infinite size. In the same way that the sphere of the moon is a mean proportional between the distance to the sun and the Earth's radius, he argued that the region of movable bodies (the mean distance of the sun to Saturn) should be the mean proportional between the distance from the sun to the fixed stars and the radius of the sun. This result is consonant with the world's three principle parts, as well as the types of motion in the universe. The sun is the primary mover, the movable bodies struggle against it, and the fixed stars surround the struggle, not moving at all. From this conception it is also appropriate that the movable bodies be seen as a mean proportional between the body that is the source of motion, and the motionless bodies that situate it. Because from the vantage point of the Earth the sun has distance $d_{se} \approx 229r_s$, and from the vantage point of the sun the distance d_{se} is approximately half the distance from sun to Saturn, the distance to Saturn is approximately 10 times the radius of the sun, or about 2,000. Therefore, the sun-stars distance is about 2,000 times the distance to Saturn, or $4{,}000{,}000r_s$, which is $60{,}000{,}000r_e$. The ancients set this distance to $20{,}000r_e$, but Kepler argued that his result is not in conflict with Brahe's observations.

Kepler was aware that the Copernican size of the universe was contentious, and argued that it is at least less unbelievable than that the sphere of the stars as set by the ancients could rotate as fast as it would need to in order to produce the apparent diurnal motion.

Once the stars are placed so far away, the question arises as to whether any of them are as great in size or luminosity as the sun. Clearly Kepler needed a negative answer to this question, because his archetypal vision required a unique sun. He argued that from the vantage point of the stars, the sun would still appear greater than any of the stars do from the vantage point of the sun. Moreover, he added, the stars appear larger than they are because of the diffusion of light.

This argument illustrates how Kepler used archetypal reasoning to access information that, given the limits on measurement precision at the time, could not be determined empirically either well or, in the case of planetary densities, at all. Lest we be distracted by the apparent strangeness of this kind of argument, or the ad hoc determination of densities, I should emphasize that this way of addressing empirically closed problems was really very clever, and the results did not conflict with the empirical evidence then available.

Although Kepler was clearly thinking in physical terms at this point, it is suggestive that he presented his archetypal arguments before the physical arguments of parts II and III. The *Epitome* was, after all, intended to bring the message of the new physical astronomy to a wide audience. Kepler seems to have thought that his physical arguments would be more persuasive if suggested in advance by an archetypal vision.

Part II

Parts II concerns the primary causes of planetary motion, by which Kepler meant the causes of annual and diurnal rotations (he addressed the cause of libration in part III). To defend the Copernican system, he needed to render physically plausible that the sun is the central body and the primary cause of the planetary motion, and that it is the Earth that rotates rather than the fixed stars. Although here it is mostly in occasional allusions that the archetypes make an explicit appearance, archetypal *reasoning* is still featured. Such reasoning plays three roles: to stand in for absent or unsatisfactory physical arguments, to link unfamiliar phenomena with the familiar, and to further support Kepler's overall world system.

Kepler's arguments that the sun is the primary cause of motion appeal to the second and third laws, the sun's beauty and general fitness for that role, and the fact that it rotates in the right direction (*Epitome*, 895; *KGW* VII, 298). In part I he argued that the sun is archetypally fit to be the central body. The argument here is that it is physically fit to be the primary cause. This mirrors the division of topics between the *Mysterium*, which addresses the topics of part I, and the *Astronomia*, which addresses those of part II. Despite the division of topics, the archetypal argument from beauty and fitness appears both here and in the *Astronomia*.

Kepler took a buckshot approach to supporting the claim that the sun is the central body of all planetary orbits, including the Earth's. He finally attacked Brahe's system directly, citing eighteen reasons to reject it. These include methodological considerations (e.g., simplicity), the archetypes (e.g., polyhedra), metaphysical considerations (e.g., because motion begins from the sun, the sun itself must be still, much like Aristotle's unmoved mover), and anthropocentric concerns (that it is to our benefit to be on a moving observatory) (*Epitome*, 908–16; *KGW* VII, 309–16). It is pretty clear from these sections that Kepler considered archetypal evidence as relevant.

Nonetheless, Kepler's buckshot approach tells us little about the connection as he saw it between the archetypes and physics. He may have wanted to persuade readers by sheer mass and variety. Perhaps too he wanted to maximize the chances that a reader would find at least some of the reasons acceptable. Of more interest is Kepler's strategy when he found his physical arguments unsatisfactory or altogether absent. It is clear that he viewed archetypal reasoning as an appropriate substitute in these cases. What justifies this view is an assumption of correspondence between the archetypal and the physical.

In the section on why it is the Earth that rotates and not the fixed stars, Kepler offered two brief arguments (having addressed the issue already in book I). First, the sphere of the stars would have to rotate unbelievably fast (*Epitome*, 887; *KGW* VII, 290–91). This argument is based on physical intuitions, and although Kepler was not completely satisfied with it, he did not say why. He offered a second, archetypal argument instead to the effect that it is not clear for whose good the stars would rotate. That it would benefit no one "destroys completely every movement of the sphere of the fixed stars" (*Epitome*, 888; *KGW* VII, 291).

The archetypes are especially important in cases where Kepler lacked a qualitative story, as in his discussion of how the sun rotates on its axis. He argued that there must be a soul in the sun to keep it rotating. This conclusion is consonant with the sun giving off light and heat, both of which are "akin to the soul" (*Epitome*, 896; *KGW* VII, 299). There is no talk of qualitative causes, only of the appropriateness of the soul to the body of the sun. Mark that Kepler distinguished between mind and soul, arguing that no mind is needed to produce the rotation. By this he meant that no intellectual faculty is required, only a living one (*Epitome*, 897; *KGW* VII, 299). This argument is clearly archetypal rather than physical.[7]

Kepler's archetypal reasoning plays an even more critical role in his physics by establishing links between different domains of phenomena. Kepler needed the link between the celestial and the terrestrial to support his argument that the primary motive force emanating from the sun is magnetic. His discussion of the nature of the sun's magnetic force and its properties echoes that in the *Astronomia*. Magnetic force, like light and heat, is a body's immaterial form, partaking of quantity, and thinning out over space. Here, however, the epistemological force of analogy is made more explicit. While the motive force is analogous to light and heat, the analogy can be used to demonstrate further properties of this force:

> These properties of light have been demonstrated in optics. The same things are proved by analogy concerning the motor power of the sun, keeping the difference between the works of illumination and movement and between the objects of each. And these same things are found to be consonant with astronomical experiments. (*Epitome*, 903; *KGW* VII, 305)[8]

This claim works on the assumption of structural correspondence between objects in one discipline and those in another. And this assumption rests on Kepler's belief that the world was created to mirror certain aesthetically pleasing ideas in God's mind. Moreover, the successful use of analogy is evidence that the relation found is one of those sufficiently pleasing for God to have replicated it.

The analogy released Kepler from the need to explain *how* the sun moves the planets. His explanation was simply a reduction to the fact that a magnet can also perform this function. We see nothing like Descartes's mechanical explanation of magnetic forces in the *Principles of Philosophy*.

There is no attempt to explain how a magnet performs this function. The fact that it does suffices. This approach is echoed in Kepler's later argument that the reason distant causes are not difficult to understand is that we have ordinary examples of such phenomena (*Epitome*, 904–5; *KGW* VII, 306).

One striking feature about the orbits is that, despite being produced by magnetic forces, the periods are regular. Indeed, this kind of celestial phenomenon inspires one to think of regulation by planetary minds. Kepler needed to show that the regular periods we observe could be produced by magnetic forces. Kepler's ideal universe was one that, once archetypally structured, could maintain this structure by material necessity alone.

He first addressed the issue qualitatively. The periods of the planets are not equal to the axial rotation of the sun because the magnetic force interacts with the planet's inertia and is spread out over the distance between the planet and the sun (*Epitome*, 899–900; *KGW* VII, 301–2). Because the magnetic force and planetary inertia are constant, the periodic times are constant, and so no mind is required to maintain the regularity (*Epitome*, 900; *KGW* VII, 302).

The next step was to address why the square of the periods is proportional to the cube of the mean distances. He identified four causes that intermingle to produce the periodic times: the length of the route traversed by the planet; the amount of matter in the planetary body (what we would call mass); the strength of the motive force, which attenuates with distance; and the amount of space the planet takes up, since a larger body will be in contact with a greater amount of the sun's motive force. (In figure 7.2 body CA is the same size as FD, but CA is twice as close and hence is in contact with twice the motive force. The same can be achieved by putting these bodies at equal distances, but decreasing the size of FD so that it is still only in contact with the motive force in area DSF.)

Earlier, in setting the densities, Kepler argued that the amount of space a body takes up is in direct proportion to the distances. He argued that because of this, the two factors balance each other out, and we need only consider the amount of matter in the body, and the length of the route traversed. The length of the route varies in direct proportion to the distance. The amount of matter is determined to be ½ power of the distances, since the volumes are directly proportional to the distances and the densities are ½ powers of the distances. Because the inertial resistance is proportional to the amount of matter, the simple ratios of the distances plus the ratio of the ½ powers yields the $^3/_2$ powers of the distances, which in turn yields the proper periodic times.

Neither physical theory nor observations laid much by way of constraints here. Even though Kepler claimed to have determined the planetary densities through archetypal arguments, these we have seen were

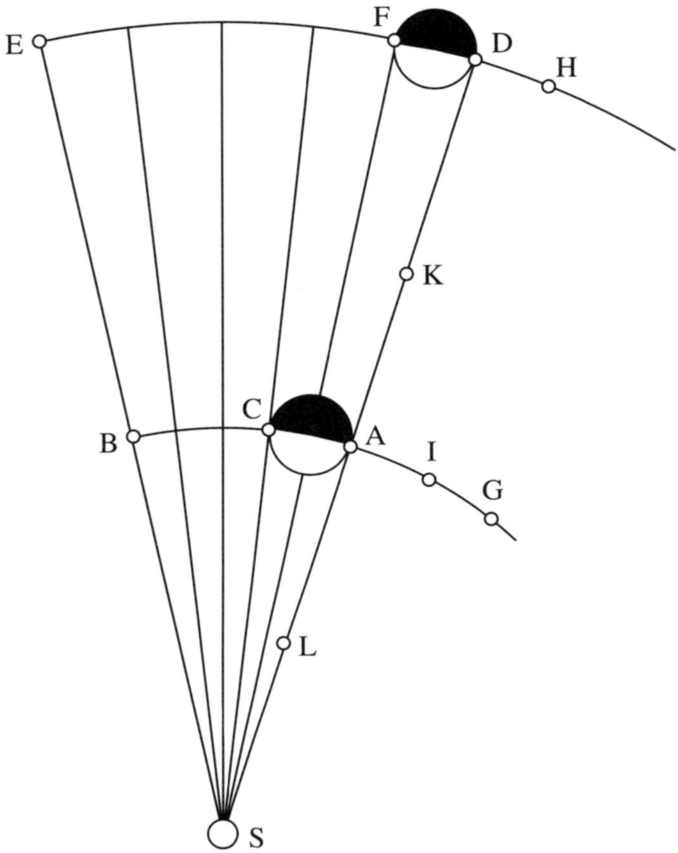

Figure 7.2. The Surface Area of the Planet and the Sun's Motive Force

unpersuasive. Kepler had neither a good physical nor a good archetypal argument to the densities of the planets. He had little to support his claim that physics could show why the third law holds. Instead he had the empirical claim that the third law does hold, and a story about why this might be. He had no persuasive independent reasons for setting the densities as he did. This underscores the point made by Gingerich (1975a, 600) and Stephenson (1987, 144) that, although the harmonic relation may be a law insofar as it was created by God, it does not support counterfactuals. In other words, because it does not follow from other physical principles, it holds only contingently for the six planets as they were created.[9]

Until now I have only alluded to Kepler's difficulty in preserving correspondence between the archetypal and the physical. As I have argued, a number of arguments in part II rely on this correspondence. Yet mixed in with these arguments already are telling signs of the tension. Just before introducing the section on the sun as magnetic cause of planetary motion, Kepler argued that we can distinguish motion produced by a mind from motion produced by a physical cause. If the motion is caused by a mind, it will produce the most aesthetically pleasing shape, whereas a physical cause need not. With regards to planetary orbits, a mind would produce regular circular motion, and so nonuniform elliptical motion "smell[s] of the nature of the balance or of material necessity" (*Epitome*, 893; *KGW* VII, 295).

This idea, that one can distinguish mental from material causation by the kind of effect produced, is new. In the *Astronomia*, the notion of mind as cause went proxy for physical theory. Nonetheless, he looked for ways to show how the mind could produce the cause (recall Kepler's discussion of how a mind could use changes in the sun's apparent diameter to determine libration). One finds no arguments in the *Astronomia* that planetary motion, given its nature, cannot be produced by a mind. Kepler's new idea was a means of determining where to look, or better, where not to look, in developing a causal theory.

As creative as this move is, and as laudable for prescinding from mental explanations of physical systems, it undercuts Kepler's general program. By divorcing the rational from the material in this way, he allowed a failure of correspondence between the physical and the archetypal. This problem occurs again in part III.

Further difficulties emerge in Kepler's discussion of the Earth's diurnal motion. He argued that 360 is the archetypal number of the revolutions of Earth, since 360 is the significant number closest to the actual number of revolutions. The discrepancy between the archetypal and actual numbers he took as evidence of two distinct causes at work: "For (1) if the number 365¼ were not composed of the two effects of two distinct causes, there would be no reason why it is not one of the archetypal numbers, that is, one of the round numbers rather than one of the disjointed and ignoble fractions" (*Epitome*, 916–17; *KGW* VII, 317).

We see a similar sort of reasoning with regards to the moon. Kepler argued that since the archetypal number of Earthly rotations a year is 360, which can be decomposed into 12 and 30, then the moon's archetypal number should fall between 12 and 30. He considered other ways of decomposing 360, but considered this the most excellent, "both geometrically . . . and harmonically, because all harmonies are represented by these two divisions of the string. Then of the numbers which, when multiplied together, give 360, there are none more beautiful" (*Epitome*, 926; *KGW* VII, 325). Kepler considered 12 the more appropriate of the two because "it would not have been of the same beauty, if there had been thirty months in the year, and each month had been of twelve days" (*Epitome*, 927; *KGW*

VII, 325). In the same way that the Earth does not respect its archetype, neither does the moon. From this Kepler concluded that, "[m]eanwhile it has been proved exactly enough in both ways [from physics and the archetypes] that this straying from the whole and beautiful numbers is due to the concurrence of the causes of the lunar movement" (*Epitome*, 928; *KGW* VII, 326).

Notice that while simple physical causes still mirror the archetypes, the actual world, as a result of the complexity of physical causes, does not. We can still infer physical causes provided the archetypes fit the data, but we have no guarantee that a physical pattern will be archetypal because it may be the product of two physical causes. This approach constitutes a new conception of what it is for the world to be in architectonic harmony with the divine blueprint. This new conception has epistemological consequences. Recall that Kepler rejected the first noncircular orbit in chapter 45 of the *Astronomia* on the basis that it is nonarchetypal. Now his claim that it can be rejected on a priori grounds no longer holds.

Kepler's methodological responses to these problems were quite ingenious. He tried to make an unfortunate turn for his worldview yield information about the kinds of causes at work (mental or physical, simple or complex). But ultimately, as I discuss in the closing chapter, these modifications undermined his program as a whole.

Part III

Part III concerns the irregularities of planetary motion and their physical causes. Once again, Kepler used the kinds of effects produced to determine whether the cause is mental or physical. This comes out particularly in his discussion of planetary libration. Kepler's first move was to point out that regular motion need not be interpreted as uniform circular motion. It could instead be taken as regular irregularities (*Epitome*, 929–30; *KGW* VII, 327–28). Indeed, regular irregularities indicate that we need to look for physical rather than mental causes. Regular nonuniform motion must be the result of "material necessity. So it is not surprising if these faculties, which are mingled together, could not attain perfection completely" (*Epitome*, 932; *KGW* VII, 331). Notice Kepler's emphasis on the inability of a physically complex world to mirror the archetypes. This emphasis is quite different from that found in the *Astronomia*, where the mathematical elegance of the versed sine measure is said to indicate, in virtue of its archetypal significance, a forthcoming physical theory.

In the *Epitome*, Kepler modified his physical theory in two significant ways. The first modification is that, because nothing on the Earth's surface could function as the attractive and repulsive poles, "we are therefore compelled to admit that the globe is inside an outer crust: in such fashion that the crust rotates during the daily movement, while the globe having the threads does not rotate" (*Epitome*, 937; *KGW* VII, 335–36). This maneuver is, of course, ad hoc. Kepler's only reasons for believing it were, first, that a two-magnet system analogous to the sun and the Earth would produce (more or less) the libratory motion and, second, that God had created a force to instantiate the archetypal construction.

As Kepler admitted in the *Astronomia*, his demonstration from the two-magnet system only shows that something like the two-magnet system might be producing the libration. We are not compelled to admit that this is the case, much less that a globe under the Earth's crust maintains the correct orientation. As a result, Kepler's archetypal arguments become very important here. Having long believed that final and efficient causes correspond, in his discussion of libration in book IV, Kepler finally made this correspondence explicit:

> The instrumental cause is the different strength of the threads, whether that is produced by nature or by posture. But the final cause is the same as that of the eccentricities themselves, namely in order that by reason of these eccentricities the movements of the planets should become very fast and very slow in such measure as would suffice for the harmonies to be exhibited through them. Book V of my *Harmonies* has to do with this. (*Epitome*, 941; *KGW* VII, 338)

> Then there will be room to speak about the final cause: to the final cause belongs the mutual tempering of the forces of libration, of the deflection of the threads, and of the revolution, in some fixed proportion . . . because the librations were prepared in order to set up the harmonies. (*Epitome*, 945; *KGW* VII, 342)

It is suggestive that Kepler emphasized his identification of a causal system in the context of his libration theory by way of its archetypal cause—an indication, I believe, that Kepler was concerned about the plausibility of his physical claims and that, once again, he brought in the archetypes to support an otherwise weak physical argument. Still, his archetypal justification suggests only that libration occurs, that it has a cause, and that the cause *might* be magnetic.

Kepler's concern is also suggested by the further methodological claim that his hypothesis is at least no less plausible than that of the solid spheres. Both hypotheses lack direct observational evidence, yet both are inferrable from the data (*Epitome*, 950; *KGW* VII, 347).

Such considerations notwithstanding, he concluded that he had shown only that there was some physical cause, and not necessarily the one he had in mind: "For between these and similar things, it is uncertain exactly what the manner of this motion is: it is alone very certain that, whatever the manner is, it has been fitted to physical and magnetic causes, i.e., corporeal

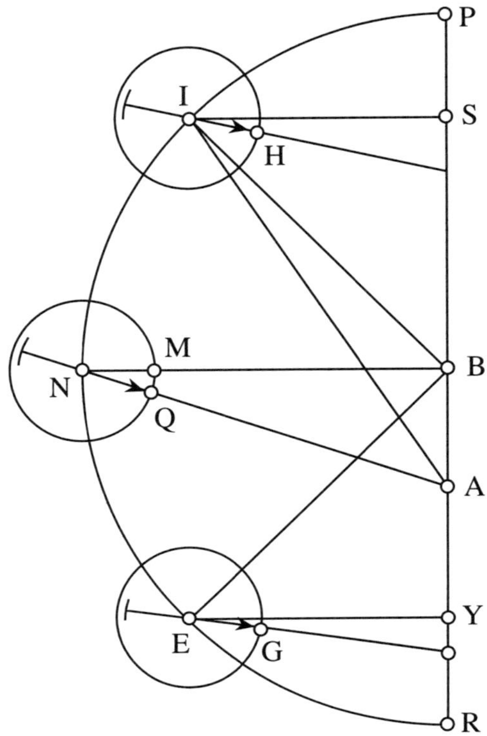

Figure 7.3. The Deflection of the Magnetic Fibers

and thus geometrical; and I have put forward examples of such causes in both causes here" (*Epitome*, 954; *KGW* VII, 350).

Kepler's second modification was to account qualitatively for how the magnetic threads of the inner crust maintain the appropriate orientation. Because the sun attracts one end of the threads and repels the others, it grabs onto them and twists them slightly (figure 7.3). As the planet is carried into the second half of the orbit, the opposite end of the thread is presented, and so the sun twists it back to its original position. To prevent the attractive pole from twisting all the way around to face the sun throughout the orbit, Kepler proposed that the twisting meets with resistance, as a compass needle resists disorientation by a magnet (*Epitome*, 942; *KGW* VII, 339–40).

If the second modification could be shown to yield precisely the right results, it would be a great success for Kepler's troubled libration theory. Despite the ad hoc nature of the inner crust hypothesis, its plausibility would be greatly increased if it could be shown that the sun, in conjunction with the natural resistance of the poles, twists them to exactly the right position. Kepler deferred this demonstration for book V, satisfied that he had shown that the world's archetypal blueprint was at least compatible with a more or less plausible, observationally adequate physics. Hence his summary comment: "[Planets] find the laws, or figure, of their movements in their own bodies, which have been conformed to Mind—not their own but the Creator's—in the very beginning of the world and attuned to effecting such movements" (*Epitome*, 960; *KGW* VII, 355).

Book V

Overall, Kepler's qualitative physics is impressive. His task in book V was to flesh out the quantitative details of his theory—to show, in other words, how mathematical astronomy can be derived from celestial physics.

Whether Kepler's mathematical theory of libration can be interpreted as a *physical* theory was a matter of some importance to Kepler, for he believed the rival theory of Fabricius, proposed years earlier, to be empirically equivalent to his own (Letter from Fabricius to Kepler, February 27, 1608; *KGW* XVI, 123–30).[10] Distressed by this empirical equivalence, Kepler criticized Fabricius's theory for its lack of physical foundations:

> You say that geometry bore you a daughter. I looked at her, she is beautiful, but she will become a very bad wench who will seduce all the men of the many daughters which mother physics has borne me. Your theory will attract lecturers and philosophers; it will offer a way out to the enemies of the physics of the sky, the patrons of ignorance. (Letter of Kepler to Fabricius, November 10, 1608; translated in Baumgardt 1951, 81; *KGW* XVI, 205. Kepler did not write to Fabricius again until October 1616)

Kepler's worry persisted, and in a letter to Maestlin in 1616 he defended his theory against Fabricius, stressing both the need to ground astronomy in physics and the plausibility of his physical theory of libration (*KGW* XVII, 201–2).

Kepler was unable to derive an adequate mathematical theory of libration from his physics, so he settled on a mathematical theory incompatible with the physical theory. Examination of this conflict reveals important methodological issues surrounding the development of nonqualitative physics.[11] Briefly, Kepler's physical theory has the sun tugging on magnetic fibers so that they twist to maintain the required orientation (figure 7.3); the sun pulls these fibers to maximum deflection at N, then pulls them back to their original position as the planet moves toward R. In book V, Kepler's mathematical account of deflection is as follows (figure 7.3). For physical reasons:

$$\frac{\sin HIS}{\sin QNB} = \frac{\sin IAP}{\sin NAP} .$$

That is, the sine of the fibers' deflection (sin HIS, sin QNB) is proportional to the sine of the coequated anomaly—a plausible explanation, given the appropriateness of using sines to measure physical forces, and the fact that the sun deflects the fibers by pulling on them. Kepler also demonstrated that the sine of the deflection is proportional to the sine of the optical equation. By the law of sines we have:

$$\frac{BI}{BA} = \frac{\sin BAI}{\sin BIA}$$

$$\frac{BI}{BA} = \frac{\sin IAP}{\sin BIA} .$$

If figure PNR is a circle, then we also have:

$$\frac{BI}{BA} = \frac{BN}{BA} = \frac{\sin BAN}{\sin BNA}$$

$$\frac{BI}{BA} = \frac{\sin NAP}{\sin BNA} ,$$

from which we derive:

$$\frac{\sin BIA}{\sin BNA} = \frac{\sin IAP}{\sin NAP}$$

Because *BIA* and *BNA* are the optical equations for the planet at *I* and *N* respectively, the sine of the optical equation is proportional to the sine of the coequated anomaly, and thus to the sine of the deflection. This account of libration was part of Kepler's formal derivation of the ellipse. Thus his mathematical account is coherent and can be derived from the physically plausible notion that the sine of the deflection of the fibers is proportional to the sine of the coequated anomaly.

The problem is that these equivalences go through on a circular orbit but not an elliptical orbit. If the sine of the deflection is set proportional to the sine of the optical equation, then on an elliptical orbit the fibers will continue to deflect past *N*. This deflection is physically implausible, for it requires the fibers to twist *against* the attractive force of the sun. Once again, Kepler's physical theory of libration was in trouble.

Without an adequate physical interpretation, the question remains whether Kepler's libration theory can be considered a part of his *physical* astronomy. As Stephenson (1987, 155) observes, that Kepler disguised these difficulties in the *Epitome*, should not be surprising because the *Epitome* was written as a textbook. Unfortunately this tactic leaves little clue as to whether Kepler viewed the problem as tractable. As I argue, however, the way Kepler introduced his libration theory in book V set the stage for a strategy that would allow him to claim that his mathematical theory of libration could still count as physical. This setup has two parts. First, he specified the conditions under which one can mathematically measure a physical force. Then he argued for an analogy between libration and the balance, the significance of which will become clear.

Kepler's second account of libration is that the measure of the net force at a given moment in time is the sine of the coequated anomaly. As discussed earlier, his motivation for this second account was that it, unlike the first, could be given an interpretation of physical significance. Here, at last, the motivation is explicitly avowed: "For since the efficient causes of the libration are the ray of the sun and the magnetic threads of the planet's body, two physical lines; it is right to seek the measure of the strength of the libration from the angle between these lines and from its sine" (*Epitome*, 969; *KGW* VII, 366). The topological features of the physical system must not only be represented but represented *appropriately* in the mathematical model. Setting libration as the versed sine of the eccentric anomaly produces the right effects, but has little to do with the structure of the physical system, because the vertex of the eccentric anomaly is at an empty point in space.

Kepler also argued that taking the sine is appropriate because: "every artificial or natural movement, in which the same or analogous principles concur, is measured out by the sines of the angles; but principally and most clearly, the movement or tendency [*nisus*] of the arms in the balance and in the lever" (*Epitome*, 970; *KGW* VII, 367). From the success in measuring other types of motion by the sines of the angles, it was natural for Kepler to draw this conclusion. After all, he believed that the physical world was created to instantiate certain mathematical relations aesthetically pleasing to God. His claim that we find motion represented in various places by the sine points to this being one of the mathematical relations that God chose to replicate.

Kepler's argument so far is that since his mathematical theory of libration has the right topological features, and since it uses a plausible means of measuring the motion, it can count as measuring the physical force that produces libration. Furthermore, the mathematical theory can count as measuring that force even without a physical theory of what causes it.

Continuing his argument, Kepler drew an analogy between libration and the balance. He did this very briefly in chapter 57 of the *Astronomia*, but here the discussion is much more detailed and explicit. First he compared the systems' structural features: "The line from the sun into the threads is like the haft of the balance, the threads like the arms of the balance; and the regions of the threads like the tray; and the weights in the trays are, in the case of the planet, the attraction towards the sun, or repulsion from the same" (*Epitome*, 970; *KGW* VII, 367). Next, he argued that the systems function in similar ways:

> For as the sun attracts the planet, so the Earth attracts bodies, and on account of this attraction, bodies are said to be heavy. . . . Accordingly that which in the balance is inequality of weights, in the planet becomes the diversity of posture of the threads with respect to the sun: for here the planet exhibits both weights on the balance as the same weight. And just as in the balance, the heavier weight comes down toward the Earth, and the lighter moves up away from the Earth; so in this case the whole globe of the planet suffers the affect [*affectionem*] of the prevailing region. . . . Therefore too the measure, in accordance with which the weights of the balance are at war with one another, will be dominant in the disbursing of this attraction and repulsion. (*Epitome*, 970; *KGW* VII, 367–68)

It is significant that in the physics of book IV the key analogy is the two-magnet system, while in the mathematical astronomy of book V the key analogy is the balance. In book IV, Kepler was arguing that the cause either is or is like the two-magnet system. In book V, he was not discussing causes at all. Rather, by explicitly detailing the balance analogy, he was arguing for the legitimacy of using methods from mechanics to analyze libration. By doing this, he tapped into an established tradition of mathematically representing physical systems.[12] Mechanics at the time was an accepted mixed science—that is, a science about the physical world that used mathematics as one of its tools. It was considered relatively unproblematic to demonstrate mathematically without reference to physical causes the properties of physical systems like the balance. Linking mechanics to celestial physics in the right way would legitimate measuring without reference to physical causes the physical forces producing celestial phenomena like libration.

Recall how Kepler's requirement in the *Apologia* that astronomy be grounded in qualitative physics allows for assessing an astronomical hypothesis on the basis of its physical plausibility. Because, due to God's creative plan, certain objects in different domains resemble each other, and because libration and the balance bear this sort of resemblance, conventional mathematical descriptions of the balance are applicable to libration theory. Consequently, since the mathematical theory has the right topological features, mirroring those of the balance, it is a physically plausible measurement of the cause—whatever it may be—of libration. Notice that the emphasis has shifted from physically plausible *mechanisms* to physically plausible *mathematical relations*. This shift is conceptually important in moving from qualitative to quantitative physics.

Concluding Remarks

The *Harmonice* shows that as a mature scientist Kepler still believed in an archetypal universe. It should now be clear from the *Epitome* that this belief was not some aesthetic hobby but intimately connected to Kepler's construction and defense of his physical astronomy. The relationship between Kepler's archetypes and his physical astronomy persisted in the *Epitome*.

The most significant role of the archetypes is to explain the design of the universe, specifically, the number, size, and spacing of planetary bodies. Such calculation allowed Kepler to access information empirically closed to him. For one example, he could not measure the diameter of the sun within acceptable margins of error. For another, he could not access empirically the densities of the planets. But he needed to access them somehow, for the only way he could account for the periods was to vary inertial resistance, and the only way he could do this without violating observational data was to vary the densities.

Structural features and physical principles, then, are both compatible with the archetypes, and interact to produce the universe as we know it. This interaction was, for Kepler, a necessary part of a complete explanation. Indeed, as we have seen, he opened his discussion of celestial physics by laying out the archetypal plan. It seems he believed his physics would be better received by a general audience if it comported with an archetypally determined universe. He used archetypal reasoning to support his physics in two ways, by linking domains of objects, and by establishing the physical plausibility of hypotheses.

These roles were quite useful in enabling Kepler to forge ahead in the virtually uncharted territory of physical astronomy. The matter, however, had become complicated in the *Harmonice* by the failure of exact correspondence between the physical and the archetypal. He responded to this problem by arguing that, unlike direct mental causes, physical causes, by their very nature, would fall short of the archetypes. This matter was further complicated by the possibility of compound causes, which would interfere with the instantiation of the archetypes. Thus did Kepler discover a physical world less perfect in representing of God's ideas than he had originally thought. Compare the following quotations, the first from the much earlier *Apologia*, the second from the *Epitome*:

> He who has learned to distinguish these things will easily disengage himself from the deluded seekers after abstract forms who quite heedlessly despise matter (the one and only thing after God), and from their importunate sophisms. (*Apologia*, 156; *KGW* XX.1, 30)

> [I]n addition to mind there was then need of natural and animal faculties also for the sake of movement; those faculties followed their own bent [*ingenium*], nor did they do everything from the dictate of mind, which they did not perceive, but they did many things from material necessity. So it is not surprising if these faculties, which are mingled together, could not attain perfection completely. (*Epitome*, 932; *KGW* VII, 331)

Notice the change in Kepler's view of matter. The first quotation chides those who fail to see its perfection, whereas the second excuses its principled imperfection. This reversal must have left Kepler disconsolate, for material imperfection mars God's communication with humanity. But the failure of exact architectonic harmony did more than threaten his sense of communicating directly with God. As I discuss in the closing chapter, it undermined much of the support he had marshaled for his physical astronomy.

Conclusion

The Fate of Kepler's Philosophical Thought

The Threat to the Perfectibility of Astronomy

The archetypes supported Kepler's realist stance in two ways. They justified his method as truth-linked, and they accounted for the accessibility to the human mind of nature's underlying structure. In particular, they justified his view that fruitfulness and simplicity are theoretical virtues on the basis of which to adjudicate rival hypotheses. Because God's essence is simple, and this simplicity is reflected in the material by archetypal correspondence, a theory that does not unify complex phenomena is unlikely to match the archetypes. As God imprinted us with the archetypes, we can rule out hypotheses that "disturbed the mind," and as the archetypes imply the unification of diverse phenomena, we have ample opportunity to test hypotheses for error. Thus, the "linking of syllogisms" implies that even if two theories are empirically equivalent, they may not be evidentially equivalent. Hence, despite making the same predictions, one theory can be better supported than the other.

Kepler's conception of the archetypal universe was also instrumental in his justification of incorporating physics into astronomy. Without this justification, his answer to the problem of observational equivalence would have been suspect for it involved the merging of traditionally distinct disciplines. Moreover, conceiving of physical objects as essentially geometrical justified the use of mathematical models in describing physical systems. The analogy between things mechanical and celestial furthered this end. The idea that some mathematical relations have more aesthetic appeal than others and hence are more likely to be found in the world justified certain mathematical models as physically plausible when no qualitative physical description was available. We see this in particular in Kepler's discussion of libration in the *Astronomia* and *Epitome*.

The archetypes also served to rule out hypotheses that could not be represented geometrically, and to establish the candidacy of those that could. Both functions were of key importance to the "war on Mars," in which Kepler broke new ground in physics and astronomy. Once he had abandoned the traditional means of representing planetary motion, he could use the archetypes to determine the plausibility of other means. In addition, the aesthetic appeal of the ellipse, its unification of his distance and area principles, recommended it over various other noncircular orbits consistent with observation.

Kepler deemed it necessary for completeness and security to provide both an a priori and an a posteriori account of the world. The a posteriori is, by its nature, based on past observations; an a priori account of the causes provides confidence that future observations will continue to follow the pattern of those past. Traditional physics and metaphysics were conducted a priori. Quantitative physics, however, was not conducted a priori, for it represented motion, as astronomy did, without reference to the causes involved. Here especially, if Kepler was to have both a priori and a posteriori accounts, he needed an archetypal model.

As already noted, Kepler argued that the nature of the material was such that it could not conform completely to an archetypal model. The problem arose during his study of celestial harmonies, and by the time he wrote book IV of the *Epitome* he had some answers. There he argued that the structure of a system could tell us something about the nature of the causes at work. A system's failure to correspond exactly to its archetype indicates a physical rather than mental cause, and more than one cause may be complicating the motion. But Kepler's earlier confidence in his method rested on the assumption of the *precision* in the physical mirroring of archetypes.

Kepler's sense of the material world's complexity was heightened by the time he finished the *Rudolphine Tables* (1627). By then he held that planetary motion is disturbed by small motions uncaptured by his planetary laws. He believed that in the orbits of Saturn, Jupiter, and Mars there were slight perturbations, which, over a great period of time, would result in a change to their orbits (Letter from Kepler to Mathias Bernegger, 20/30 June 1625; *KGW* XVIII, 237).[1] He came to believe that the moon's orbit and the rotations of the sun and Earth were also slightly irregular, which in turn raised doubts about the regularity of the inner planets.

Although the perturbations were small, Kepler spoke of them as extraordinary, disturbing the mathematical regularity expected of planetary motions.[2]

In commenting on Kepler's response to the planetary perturbations, Wilson argues that: "The complex interaction of physical influences, and the complex artistry of the Creator in adapting his creation to several harmonic patterns simultaneously, could leave the exactitude of any particular relation in doubt" (Wilson 1978, 241). While the complexity of physical and archetypal causes may complicate theorizing, this by itself does not threaten the regularity of mathematical descriptions of physical relations. Rather, the threat comes from the misbehavior of matter, the "intensiones et remissiones recipientium extra ordinem." This interpretation suggests that Kepler did not locate the problem in the interaction of physical causes and the complex artistry of the Creator. The problem lay rather in the break between the material and the archetypal, a break consequently that threatened the perfectibility of astronomy.[3]

This threat had serious consequences. If matter is not fully tractable to archetypal rule, then Kepler's view of planetary orbits loses an appreciable amount of the theoretical support he thought he had adduced. For example, the application of mechanics to libration rests on the assumption that certain mathematical relations in mechanics can describe the motion of the planets exactly (it is certainly not supported by the assumption of the similarity of the nature of the causes since all the parts of the standard machines analyzed in mechanics are in contact with each other, whereas the sun acts on the planets at a distance). Without this assumption of exactness, the planet could deviate slightly from its predicted path, and given the problem of observational error, Kepler would not be the wiser. This possibility jeopardizes his choice of the ellipse over other figures equally compatible with observation. The unification of the distance and area rules count no longer as evidence for but merely as a pragmatic virtue of the ellipse. Kepler's realism is defunct.

Not only does Kepler's view lose the theoretical support on which his realism depends, the now possible irregularity of the physical world places predictive confidence in jeopardy. Kepler's laws need not persist in holding if the material is wont to misbehave. It is interesting how in this case the failure of archetypal correspondence endangers the predictive adequacy necessary, but clearly not sufficient, for a theory's vindication.

Because the material may fail to match the archetypes precisely, if these perturbations have physical causes (and surely for Kepler they do), then a complete physical theory no longer maps the archetypes. Formal, final, and efficient causes fail to correspond, and so the archetypes cannot be used to rule out certain physical hypotheses as implausible. Furthermore, because it was the archetypal correspondence of physics and astronomy that justified their merger, that merger must now be justified on other grounds. For the same reason, archetypal reasoning can no longer make plausible physical models unsecured by qualitative physics.

If the nature of matter precludes astronomy from being an accurate science, Kepler's earlier emphasis on precision is of little use. Small, observed, deviations above tolerance from the results predicted by a theory need not in any way affect its assessment, and the requirement of a complete and precise archetypal construction is now obsolete, since there may not be one to be had.

Nonetheless, these problems do not diminish but rather highlight the historical significance of Kepler's archetypal universe. This vision inspired and, for a while at least, justified the discoveries and methodological innovations that earned him a place of high distinction in the scientific revolution. He envisioned a universe perfectly created in the perfect image of order, an order at once in consort with and knowable by the human mind. Perhaps this is why those who study Kepler's archetypes tend to speak with an air of regret that such views can no longer be expected to be part of scientific inquiry. As Caspar puts it; "Certainly, with the progress of knowledge, some of Kepler's views have proved erroneous. His view of nature was conditioned by the time. But the ethos by which he performed his work [that nature can be understood, not just cataloged] has also changed and this is a pity" (Caspar [1959] 1993, 387).

Kepler's Philosophical Legacy

I would like to be able to say that Kepler's philosophical legacy lives on, that he is the father of modern philosophy of science just as Descartes is the father of modern epistemology. But this is not so. Kepler shows up frequently enough in philosophy of science debates but seldom as a philosopher. I would like to explain this neglect, but the task is a difficult one. After all, if Kepler's philosophy has not been discussed, history will indicate at best a series of conspicuous absences.[4] Two such dearths are in the Descartes-Mersenne correspondence, and the Leibniz-Clarke correspondence.[5]

Kepler's philosophy, however, was discussed by his contemporaries. The *Mysterium* garnered enough interest to warrant a second edition. His work on harmonies was appreciated by Mersenne and Horrocks and challenged by Fludd. Also controversial were his archetypal justification of grounding astronomy in physics, and applying mathematics to physics. These views received largely negative responses from Brengger, Maestlin, Crüger, Boulliau, Riccioli, and Longomontanus (Applebaum 1996, 459, 499). My interest, however, is less whether Kepler's views had currency at the time, and more why they have none in the history of philosophy today. This question is particularly keen considering both the sophistication and contemporary appeal of certain key features of Kepler's philosophy—hence the following focus on well-studied sources where one might reasonably expect Kepler's philosophical ideas to appear.

Marin Mersenne was a key intellectual figure of the mid-seventeenth century. He knew of and corresponded with many of the important thinkers of his time, most notably Descartes and Huygens, and his collected correspondences provide a valuable resource for studying the transmission of ideas at that time. Mersenne was aware of not only Kepler's astronomy but also his work on world harmonies. Indeed, Kepler complained that Mersenne was dragging him into a debate with Fludd (Letter to Jakob Bartsch, Novemebr 6, 1629; translated by Baumgardt 1951, 178). Mersenne clearly felt that Kepler's work was worth writing about, and spoke often of him in letters. Kepler's magnetic philosophy was a topic of great interest; the central controversy was over the possibility of a cause acting at a distance from its object. Citing the *Astronomia*, Deschamps reported to Mersenne that he approved of the magnetic philosophy of Kepler and Gilbert (December 1640; Tannery 1932–88, X, 123). Descartes, in contrast, argued that all causes are contact causes, and that consequently the planets indubitably move through a heavenly fluid (April 1646; Tannery 1932–88, XIV, 227). Jacques de Volois approved of Boulliau's method of calculating the ellipse without requiring physical considerations (June 1648; Tannery 1932–88, XVI, 376–77). Kepler's optics was also discussed, in particular by Descartes.

Given Descartes's awareness of Kepler's work, and certain philosophical affinities, one would expect Descartes to have discussed Kepler's philosophy explicitly in these letters. But he did not, although curiously he wrote to Mersenne that a priori knowledge of the heavens would be preferable to knowledge derived solely a posteriori. He lamented the lack of such an account, but made no mention of Kepler's attempt to provide one (May 1632; translated by Kenny 1970, 23–24). In a letter to Vatier he discusses his a priori method in a manner strikingly similar to Kepler's; that hypotheses should first be deduced a priori from first principles, and then supported a posteriori (February 1638; translated by Kenny 1970, 48). Both Kepler and Descartes saw physical objects as essentially geometrical, and both appeal to this notion to justify the use of mathematical mechanics in physics. In the third *Meditation*, Descartes claimed that our idea of God is formally caused by an archetype, namely, the essence of God.[6] Although Kepler was not referred to in these contexts, there is at least some evidence that his influence on Descartes was not negligible. Consider Leibniz's claim that Descartes employed some of Kepler's physical results "brilliantly, although, as is his custom, he concealed their author" (Applebaum 1996, 454; quoted from "Tentamen de motuum coelestium causis," *Acta eruditorum*, February 1689, by Cohen 1972).

Leibniz himself was very familiar with Kepler's work, not just through hearsay, and not just the astronomy and optics. Leibniz aided Michael Gottlieb Hansch in his efforts to publish Kepler's manuscripts and so had intimate contact with Kepler's work (Caspar [1959] 1993, 366). Also, there is evidence suggesting that Kepler influenced Leibniz. Like Kepler, Leibniz believed that everything was created for a reason (Leibniz codified this as the principle of sufficient reason). Both also believed that the universe, once created, would function beautifully in accordance with physical laws. Leibniz (1962, 129), in a very Keplerian passage, claimed that as a result of divine providence, the physical world would express aesthetically pleasing geometry and harmony. He believed, famously, that God was logically constrained to create not a perfect world per se, but the best of all possible worlds. Kepler had arrived at this conclusion earlier after his struggle to fit harmonies to planetary motion. There is also direct evidence that Kepler influenced Leibniz. He explicitly credited Kepler with discovering how nature instantiates the harmonious laws of God. Meli observes:

> In the "Tentamen" Leibniz praised Kepler for having first made known to the mortals "*jura poli, rerumque fidem, legesque Deorum*." This quotation from Claudian, which originally referred to Archimedes, mentions three main areas in which Kepler was important for Leibniz, namely astronomy and the laws of planetary motion; the order, regularity and especially harmony of nature; and the role of theology in many aspects of his work. (Meli 1991, 424)

This passage appears in Leibniz's reproach to Newton for not grounding his mathematical physics in qualitative physical theory (Meli 1991, 424), a matter that worried Kepler. This influence on Leibniz is not widely known; it has taken the work of scholars like Meli to bring to light Kepler's influence on Leibniz,[7] probably because his influence qua philosopher is not readily apparent in Leibniz's widely read works.

One notable absence is from the well-studied Leibniz-Clarke correspondence (written in 1715 and 1716). Leibniz had been embroiled in a conflict with Newton and his followers for several years, and Clarke stepped in to answer Leibniz's objections to the religious and philosophical aspects of Newton's work. The debate ranged from the relationship between God and the created world, to the perfection or imperfection of matter, and to the need to explain why God created the universe this way rather than some other. These are topics to which a discussion of Kepler's philosophy might prove germane, yet again, when Kepler is mentioned it is only in the guise of scientist (e.g., Alexander 1956, 88, 135).

Interpreting these conspicuous absences is difficult, especially as it was common enough at the time for philosophers to refrain from citing sources. While it is clear that Kepler's philosophical views are of considerable interest, he seems a bit of a "lost philosopher" whom I have worked to bring to light. But the question of whether he was a philosopher of *influence* has barely been broached, much less more involved questions about the nature and course of such possible influence. Were subsequent approaches to scientific method, such as Whewell's, influenced by Kepler? Were apparent influences the indirect result of a wider trend? Was Kepler's influence due to his method as practiced in astronomy and optics, his philosophical remarks on method, or both? As difficult as these questions are, answering them would be no less interesting and rewarding.

Whether or not it influenced Descartes, and even though it influenced Leibniz, Kepler's philosophy was marked for extinction. It is ironic that what makes his work a stimulant to philosophy of science may be precisely what sealed its historical extinction. With the exception of the *Apologia*,

Kepler's philosophy has to be sifted from his discussion of the archetypes, from his astronomy, and from his optics. His requirement of the archetypes that they be both precise and testable renders them fascinating but obsolete. His achievements in astronomy and optics are so great that they cannot help but overshadow Kepler's methodological commitments, even though, as I have argued, these commitments made his remarkable discoveries possible. Nonetheless, it is this blend that reveals the full importance of Kepler's philosophy, for it illustrates beautifully how philosophy can and did contribute to the improvement of an empirical discipline.

Notes

Introduction

1. George Smith's phrase, in Harper and Smith 1995, 147.

2. Kepler's dedicatory letter in the *Astronomia* to "Rudolph II, The Ever August Emperor of the Romans, King of Germany, Hungary, Bohemia, &c. Archduke of Austria, &c." (*AN*, 30; *KGW* III, 7) plays out the metaphor that his study of Mar's orbit was a war ("I myself shall occupy myself with Astronomy, and, riding in the triumphal car, will display the remaining glories of our captive that are particularly known to me, as well as all the aspects of the war, both in its waging and in its conclusion"; *AN*, 31; *KGW* III, 8).

Chapter 1
Kepler's Life and Times

1. Duke Ulrich had established a scholarship system designed to aid "children of poor, pious people, of hard-working, Christian, God-fearing character and background, and suited for study" of theology (Methuen 1998, 37).

2. Some of Maestlin's astronomical views are discussed in the next section. See also Jarrell 1971 and Westman 1975 on Maestlin.

3. This comment suggests that, although Kepler was aware (indirectly through Melanchthon and directly through Maestlin; Methuen 1998, 74–78, 155–58) of the idea that astronomy could reveal God, he needed to prove this connection to himself.

4. See Thoren 1990, 432–39; Jardine 1984, especially ch. 1; Rosen 1986; and Gingerich and Westman 1988, 42–76).

5. I have not intended to provide a comprehensive biography of Kepler. Rather, I have simply sketched certain key events surrounding the writing and publication of the works discussed here. For a detailed and moving biography, see Max Caspar's ([1959] 1993) definitive *Kepler*.

6. Recently Yavetz (1998, 221–78) challenged Schiaparelli's 1877 interpretation of Eudoxus.

7. The author of *Mechanics* is unknown, but is generally thought to be an early Aristotelian, perhaps a pupil of Aristotle.

8. Moreover, because physics was the study of natural motion, and the experimental method of manipulating objects diverts them from their natural motion, such a method would seem inappropriate.

9. The details of how Ptolemy used these devices to develop an incredibly accurate predictive astronomy are well beyond the scope of this book. See Neugebauer [1957] 1969, and Toomer's (1984) translation of the *Almagest*.

10. Francesco Patrizi (1529–97), who believed in heavenly fluidity, was mocked frequently by Kepler (*Apologia*, 155; KGW XX.1, 29; *AN*, 117; *KGW* III, 62).

Despite Kepler's mocking tone, he was raising a serious concern—namely, that no explanation has been offered of how the planets move through the fluid.

11. For further reading on Copernicus, see Swerdlow 1973; Swerdlow and Neugebauer 1984; Copernicus's *De revolutionibus* (translated by Rosen 1992).

12. Dreyer's ([1906] 1953, 345–63), Kuhn's (1957, 185–209), Koyré's ([1961] 1973, 55–66), and Cohen's ([1960] 1985, 45–52) popular books have been instrumental in bringing to our attention the reasonableness of the resistance to Copernicanism.

13. I am setting aside the debate over whether Aristotle's methodological prescription was intended to apply to scientific discovery, or the presentation of results; see Barnes [1957, 1965] 1975, 65–87.

14. It is worth noting that astrology was viewed by many as a practical discipline with political import (including Kepler, see *HM*, bk. 3).

15. See Clagett [1959] 1961; Drake and Drabkin 1969, introduction; and Rose and Drake 1971.

16. Kepler was certainly not the first to speak of the world as a machine; for example, Nicole Oresme (1323?–1382) used the metaphor of the world as a mechanical clock (Wallace 1981, 42).

17. See Conger [1922] 1967, for a survey of Neoplatonism in the Renaissance. See also Westman 1972 and Kristeller [1943] 1964; 1979.

18. By 1620 Kepler had heard of Bacon, and Bacon of Kepler through Henry Wotton (there is a letter from Wotton to Bacon in *KGW* XVIII, 42). But by this time Kepler already had a well-developed conception of the way the mind should process data.

19. This letter appears in *Tychonis Brahe Opera Omnia*, VIII, 52–55; and is reprinted in *KGW* XIII, 204–5.

Chapter 2
The *Mysterium cosmographicum* and Kepler's Early Approach to Natural Philosophy

1. In 1621 Kepler published a second edition of the *Mysterium*. He left the original unchanged, but added notes to clarify, revise, and, in some cases, recant his original views. Unless otherwise stated, the references to the *Mysterium* are to Kepler's original publication in 1596.

2. Despite Kepler's use of Aristotelian language, he did not, to my knowledge, discuss material causes.What is important here, however, is how Kepler used these terms, not whether that use was faithful to Aristotle.

3. "In globo igitur est trinitas. . . . Sex igitur mobilia Solem circumambeuntia. . . . Sic igitur Sol in medio mobilium quietus ipse et tamen fons motus gerit imaginem Dej patris creatoris. Nam quod est deo creatio, hoc est Solj motus. Movet autem in fixis, ut pater in filio creat. Fixea enim nisi locum praeberent sua quiete, nihil moverj posset."

4. One need only check the *Personenregister* in *KGW* XIII–XVIII to confirm this.

5. See Field 1988, 1–16, for a comparison of the use of the polyhedra by Plato and Kepler.

6. Kepler did not engage in such speculations (that the icosahedron is most like

the sphere, or that the order from sun to stars should be from most important to least), so as a result we are not offered an explanation on a priori grounds of why these alternative orderings are inappropriate.

7. Here Kepler was concerned with ordering the solids on the basis of relative distances between the planets. He later tested the polyhedral hypothesis more rigorously.

8. Kepler corresponded frequently with Hans Georg Herwart von Hohenburg, who was the Bavarian chancellor, on both religious and scientific matters. Despite their religious differences (Herwart was Catholic, and Kepler was an eclectic Lutheran with sympathies for both Calvinist and Catholic doctrines), Herwart was a supportive patron of Kepler's scientific work (Caspar [1959] 1993, 80–85).

9. This result probably did not matter much since Kepler was already convinced of the correctness of his polyhedral model before getting the new values from Maestlin, as he indicated in a letter to Maestlin written in 1595 (Grafton 1973, 524).

10. That is, using base 60 rather than base 10. In base 60, 60 seconds make up 1 minute, 60 minutes make up 1 degree.

11. These remarks suggest that Kepler may not yet have been aware that the preface to *De revolutionibus* was written by Andreas Osiander, not by Copernicus himself. He may have been aware of this, however, since Maestlin suspected as much (Westman 1975, 331). Kepler was certainly cognizant of this by the time he wrote the *Apologia* (*Apologia*, 150; *KGW* XX.1, 26–27).

12. More precisely, Kepler claimed that geometrical figures could not inhibit motion, "For no point or center has weight" (*MC*, 167; *KGW* I, 56).

13. Acronychal observations are made when the Earth is in line with the sun and the observed body. When the model is based on the mean sun, then the observations are taken when the mean sun is in line with the planets.

14. In the *Apologia* Kepler cites William Gilbert's work on magnets as filling a gap in the *Mysterium* (*Apologia*, 146; *KGW* XX.1, 24). I found no evidence that Kepler was aware of this gap while writing the *Mysterium*.

15. This quotation is from Kepler's annotated notes in 1621. Kepler saw himself as expressing more perspicuously what he already believed in 1596. My reading of chapter 11 is in agreement with this self-assessment.

16. This term is Jardine's (1984, 252). See also Westman's (1972, 233–64) discussion. Westman uses the term "microcosmic harmony" to refer to the harmony between the different levels.

17. We see this view exemplified in Marsilio Ficino (1433–99), Pico della Mirandola (1463–94), Robert Fludd (1574–1637), and others. See Conger 1967, for a survey of Neoplatonists committed to architectonic harmony. See also Westman 1972 and Kristeller [1943] 1964; 1979.

18. Fabricius, a well-known astronomer with whom Kepler corresponded frequently, was baffled by Kepler's insistence on using physical considerations in developing astronomical models.

19. Kepler discussed this issue in chapter 21 of the *Astronomia* and in the *Apologia*. Bennett (1998) explores this principle as it is presented in the *Apologia*.

20. Lakatos, a twentieth-century philosopher of science, proposes that one can adjudicate between hypotheses on the grounds of whether they make novel predic-

tions. The classic example of this is the prediction of the existence of Neptune (independently) by Leverrier and Adams. They observed that the deviations of the orbit of Uranus could be explained by appeal to the existence of another planet influencing its orbit. The subsequent discovery of Neptune was a striking success for Newtonian mechanics. Zahar suggests that requiring a progressive hypothesis to make novel predictions is too strong. He recommends modifying the requirement to include hypotheses that provide novel explanations, that is, explanations of known phenomena that the hypothesis was not constructed to explain (Lakatos and Zahar [1976] 1978, 184–85).

21. This proviso reminds one of the modern notion of falsifiability.

22. He even considered the possibility of inserting invisible planets between the visible ones for the purpose of ordering the ratios of the distances numerically (*MC*, 63–65; *KGW* I, 10).

Chapter 3
Kepler's *Apologia*: An Early Modern Treatise on Realism

1. The *Apologia* appears translated in Jardine 1984.

2. It should be noted that Kepler was concerned not with general philosophical arguments for skepticism, but rather with addressing problems specifically astronomical. His response to these is nonetheless of general philosophical interest.

3. The term "realism" is a modern one which Kepler did not use. I am using it here to refer to Kepler's belief that with the appropriate method one can access truth.

4. Although Jardine (1979; 1984) and Gardner (1983) have commented on the modern appearance of the *Apologia*, and both noted Kepler's belief in a world created to be epistemically accessible, my purpose is to show explicitly how Kepler's cosmology justified his response to the skeptics.

5. See Jardine 1984; Gingerich and Westman 1988; and Rosen 1986, for a discussion of Brahe's priority dispute.

6. Kepler pointed out that even if one is indifferent to the physical implications of hypotheses, it is not as easy as Ursus supposed to construct a variety of empirically adequate observationally equivalent hypotheses (*Apologia*, 147; *KGW* XX.1, 25).

7. Even Maestlin, Kepler's teacher, was puzzled by Kepler's use of physics in astronomy. See Maestlin's letter to Kepler, September 21, 1616, *KGW* XVII, 186–88.

8. Jardine (1984) observes that Tycho saw this as a misinterpretation of Patricius (*Apologia*, 155).

9. *De rebus naturalibus* . . . (Cologne, 1590), *De natura coeli*, ch. 6, col. 252; cited in Donahue's introduction to his translation of Kepler's *Astronomia*, 6.

10. "[A]lmost all notable authors to date assess them [astronomical hypotheses] on both geometrical and physical grounds and want them to be confirmed in all respects" (*Apologia*, 140; *KGW* XX.1, 21).

11. Harper and Smith, in their study of Newton, emphasize the substantial nature of the support provided by using diverse phenomena to measure the same parameter of a theory that is used to explain those phenomena. The Copernican hypothesis "turns data . . . into evidence" (Harper and Smith 1995, 147) by unify-

ing planetary models, thereby allowing each retrograde motion to be a function of the Earth's orbit (Harper and Smith 1995, 146–48). (Harper and Smith's term "systematic dependencies" is roughly analogous to "linked syllogisms.") It is important to note that, as interpreted, Kepler's notion of the linked syllogisms is stronger than that of falsifiability. Simple contingent assertions like "there is a table in this room" are falsifiable, but do not involve a linking of syllogisms. Such simple contingent assertions entail two types of predictions: that when we look, it is true that we will see a table, and when we look it is false that we will not see a table. If a prediction is generated by linked syllogisms, by contrast, then it is strongly entailed by them. A prediction is weakly entailed by a theory if, when falsified, the required modifications to the theory are not extensive. A prediction is strongly entailed if, when falsified, the required modifications are extensive. So, for example, if under the Ptolemaic system the retrograde motions of Saturn occur earlier than expected, then the model for Saturn can be modified without requiring any modifications of the models for the other planets. If such a change occurs under the Copernican system, the matter is not so simple, and may require modifications to the models for the other planets as well.

Chapter 4
Kepler's Archetypes and the *Astronomia nova*

1. Originally titled *Astronomia Nova ΑΙΤΙΟΛΟΓΗΤΟΣ, Sev Physica Coelestis, Tradita Commentariis De Motibus Stellæ Martis Ex observationibus G. V. Tychonis Brahe*. Donahue translates the title as "New Astronomy Based upon Causes or Celestial Physics, treated by means of commentaries on the motions of the star Mars from the observations of Tycho Brahe, Gent." (*AN*, 27). See Stephenson 1987 for a detailed account of the role Kepler's physics played in the *Astronomia*. See also Small [1804] 1963; Russell 1964; Wilson 1968; 1972; Aiton 1969; 1973; Gingerich 1973; Whiteside 1974; and Beer and Beer 1975.

2. He offered this kind of justification of the ellipse later, in the *Harmonice*.

3. I do not, however, follow the order of Kepler's arguments that the mean sun should be replaced with the true sun, since his arguments for this span the better part of the *Astronomia*. These arguments are discussed together. I also do not offer a thorough discussion of some important topics in the *Astronomia*, like his solar theory or his investigation of latitudes in part V, though some of the results of these topics will be touched on where appropriate.

4. One wonders whether at this point Kepler considered his polyhedral hypothesis to be only probable, because the polyhedra are not directly observable. Archetypal hypotheses have an advantage over physical ones, however, because for Kepler they are directly imprinted on the human mind.

5. These arguments are in chapters 1–6, 12–14, 52, 61–62, 67.

6. "$\odot$" refers to the true sun, "$\bar{\odot}$" to the mean sun.

7. Acronychal observations are made as if from either the mean or the true sun and are accomplished by waiting until the observed position of the planet is on the opposite side of the Earth from the true or mean sun.

8. It is not clear when chapter 52 was written. Stephenson (1987, 106) argues that it was written well before chapter 53, that it was one of Kepler's earlier results. Even so, given the order in which Kepler presented his results, he expected his

readers to be sufficiently persuaded by the end of chapter 6 that exploring models based on the true sun was a worthwhile project.

9. Notice again that Kepler considered his physical reasoning to be a priori rather than a posteriori.

10. At various points in the *Astronomia*, Kepler considered the possibility of whether a mind moved the planet. This view may seem at odds with his commitment to ground astronomy in physics, but it is not so. For Kepler, minds could count as ordinary causes of motion since, after all, we often experience the causal relations between our own minds and bodies. He was very careful not to attribute any powers to planetary minds that we ourselves do not possess. The planetary minds could not simply determine where the planets are supposed to be, but rather, required some means of calculating these positions. In addition, Kepler used the notion of planetary minds to rule out certain kinds of causal patterns as implausible. If a mind cannot calculate its proper motion, then no physical force can produce that motion. Because all motions are created by God, who is a mind, then the motion must be calculable for God to have created it.

11. The situation is worse for an epicycle model based on the mean sun because there are two empty points that need to be causally efficacious.

12. That is, showing that the lines of the nodes of Mars pass through the sun, and that the inclinations of orbital planes do not wobble if they pass through the sun.

13. Gingerich [1964] 1993, 219) suggests that one should not take too seriously Kepler's claim to have required until the fifth year to complete these calculations. Gingerich offers two accounts of the number of trials required for a computer to carry out Kepler's calculation of the eccentricities of the vicarious hypothesis (nine rather than Kepler's seventy are required; Gingerich [1964] 1993; 1973. In the later of these two papers, Gingerich concludes that Kepler repeated the calculations on redundant data in an attempt to ensure that problems with the observations would not infect the construction of the hypothesis.

14. In chapters 10–15 Kepler worked on recalculating and justifying these calculations of Brahe's observations in order to get positions as they would be seen from the true rather than mean sun.

15. Observed by Bennett (1998).

16. Kepler was not the first to try an unequally divided eccentricity. Tycho Brahe also used a nonbisected eccentricity, as Kepler noted (*AN*, 251; *KGW* III, 152–53).

17. Most of Kepler's physical arguments occur in his section on solar theory, which I discuss in abstraction from his solar theory.

18. Notice that Kepler did not claim that the universe is completely comprehensible by us.

19. Kepler spoke in these terms in the *Epitome* (904–5; *KGW* VII, 306), but it should become clear that he already had something similar in mind when writing the *Astronomia*.

20. In the *Epitome*, Kepler argued that although the concept of a (physical) motive force is less intuitive than the concept of planetary movers, the former is still comprehensible, and that is what counts: "Their being more difficult to believe is no disadvantage, provided they are easy to comprehend; . . . And the mode by

which we perceive in mind what sort of form it is does not disturb us: alone the unbelievable strength [*fortitudo*] of this form keeps us in doubt" (*Epitome*, 904–5; *KGW* VII, 306).

In the *Harmonice*, he made explicit the connection between comprehensibility and the archetypes: "For if they are knowable, then they can enter the Mind and into the shaping of the archetype; but if they are unknowable (in the sense which has been explained in Book I) then they have remained outside the Mind of the eternal Craftsman, and have in no way matched the archetype" (*HM* 139; *KGW* VI, 100–101). In this context, "knowability" has a more technical geometrical meaning. A knowable figure is constructable from either the diameter of a circle or the square of that diameter.

21. Later in the *Astronomia*, Kepler proposed that with a noncircular orbit, the measure of libration did form a simple proportion with the changes in the apparent diameter of the sun, and so could be used by a planetary mind to determine where it needed to be (*AN*, 565–67; *KGW* III, 359–61; see below).

22. *Ad Vitellionem paralipomena, quibus astronomiae pars optica traditur*, in *KGW* II. Kepler began work on *Astronomiae pars optica* in 1600. Although it is not discussed here, this book marks another of Kepler's great achievements. It is a thorough exploration of optics, from the metaphysics of light, to astronomical optics, to how it is that the eye receives images. Kepler was the first to observe that the retinal image is inverted. See Straker 1981, for a discussion of Kepler's work in optics.

23. Kozhamthadam (1994, 21–22) makes a similar point. See Lindberg 1986, 5–42, for a more detailed discussion of the Neoplatonic influence on Kepler's optics.

24. Stephenson (1987, 69–75) was the first to make clear that the simple inverse proportion of the attenuation of the sun's motive force was not due to its traveling out only on the ecliptic (Kepler believed that it spread out in all directions), but because on the basis of the analogy with light, Kepler believed that it took the form of an *image* of the sun's motion. This image moves in a circle, as the sun spins, and it loses its power over distance in proportion to the distance from the sun.

25. Kepler's discussion of the second law took place in his examination of the Earth's orbit. Because he clearly saw the second law as applicable to Mars, details about the Earth will be left out of my discussion.

26. Although in some sense the circle is still prior in Kepler's system, because the sun pushes the planets in a circle, it is the planets that disrupt the circular motion with their libration.

27. In the *Mysterium* Kepler used the phrase "lucky guess" to refer to Copernicus's a posteriori argument for his hypothesis, and suggested that one also needed a priori justification from the causes (both physical and metaphysical).

28. By accepting methods of mechanics, Kepler was not committed to mechanical philosophy. The method of Archimedean mechanics involved the reduction of the physical properties of a system to its geometrical properties for the purposes of mathematical analysis. Mechanical philosophy, by contrast, is the position that all causes are contact causes. See Chapter 5.

29. The system will not actually be in equilibrium because the transverse pushing force exerted by the sun will constantly be changing the direction of BK by

moving the planet to another position. If the system is considered without the transverse pushing force, however, the direction of BK will not change, and so Kepler argued that at a given instant the system will fall into equilibrium.

30. Kepler made it sound as though once the scales fell from his eyes, the ellipse would obviously follow from his account of libration: "O ridiculous me! To think that the reciprocation [libration] on the diameter could not be the way to the ellipse! So it came to me as no small revelation that through the reciprocation an ellipse was generated" (*AN*, 576; *KGW* III, 366). It was not that easy, however. In chapter 58 he interpreted his libration theory to imply a different placement of the planet, producing an ovoid orbit. He called this orbit the *via buccosa*, since it turned out to be too "puffy-cheeked." Stephenson (1987, 122–26) offers a nice description of the problem and its connection to the difficulties in interpreting the eccentric anomaly under a noncircular orbit.

Chapter 5
The Aristotelian Kepler

1. See Methuen's 1998 study of the curriculum at Tübingen, and Schmitt 1975, for a more general discussion of university curricula during the sixteenth and seventeenth centuries.

2. One should bear in mind that Aristotelianism in the Renaissance was often blended with elements from Neoplatonism, Christianity, and other traditions. See Schmitt 1983; Wallace 1988; and Kristeller 1979.

3. See Jardine 1984, 232–41, for a finer grained analysis of responses to the conflict between astronomy and Aristotelian physics.

4. Pereira referred to astronomers as "astrologers," for at that time astronomers typically studied astrology as well. Kepler himself constructed numerous astrological calenders.

5. From Pereira, *De communibus omnium rerum naturalium principiis & affectionibus* ([1562], 1576, 47D-48B; cited in Jardine 1984, 238).

6. Wilson (1974, 258) observes that Kepler's "forces" were mechanical not in the sense that they consist of bodies bumping together (the primary motive force was an emanation from the sun), but in the sense rather that they are subject to the principles of mechanics.

7. Modern commentators dispute whether Aristotle's "injunction" was intended as an outright ban, or merely a mild warning. For the former interpretation, see Kozhamthadam 1994, 57, 275; for the latter, see Drake 1975, 239.

8. It seems almost inconceivable today to think of mathematics as foreign to physics. But in Kepler's time it seemed almost inconceivable to analyze physical forces mathematically. Johann Brengger, for example, expressed his bafflement to Kepler on this matter (October 30, 1607; *KGW* XVI, 71).

9. As mentioned earlier, Kepler pointed out that "William Gilbert the Englishman appears to have made good what was lacking in my arguments on Copernicus's behalf through his admirable skill and his industry in collecting observations in the study of magnets" (*Apologia*, 146; *KGW* XX.1, 24).

10. Kepler reported that he was familiar with this work (*KGW* XIX, 328–29).

11. For Aristotle, the higher discipline will have fewer principles, which parallels Kepler's conviction that the correct archetypal explanation will be simple, unifying diverse physical phenomena. As noted earlier, Kepler appealed to this conviction in motivating the use of diverse phenomena to measure the same parameter.

12. Westman bases his translation of this passage on Strong's (1966, 173).

13. I suspect there was ironic satisfaction in Kepler's criticism. In the notes annotated to the *Mysterium* he observed that Aristotle "carped [at the idea that mathematical objects could cause physical motion] in so many places" (*quod dogma Aristoteles tot locis vellicauit*; *MC*, 125; *KGW* VIII, 62).

14. This interpretation, that the final and formal features of all disciplines are unified, only justifies the view of terrestrial physics *as analogous* to the celestial. To view celestial forces as the same as terrestrial is to see the efficient causes as unified.

15. Aristotle listed mechanics as a discipline to which geometrical demonstrations could be legitimately applied (*Posterior Analytics* 76a23–25). For a discussion of the controversy over the relationship between mathematics and mechanics during the Renaissance, see Drake and Drabkin 1969; Rose and Drake 1971, 65–104. There is good reason to think that Kepler's reduction of libration to the balance or lever was a reduction to Archimedean statics. He treated the system as though it were in equilibrium, proceeding from there in the standard Archimedean manner.

16. Kepler tended to speak of the measure of libratory force as the "strength of an angle."

17. Notice the distinctly Aristotelian interpretation of a Platonic cosmology. The view is that quantity is separable from matter only through the act of mental abstraction (Aristotle's *Physics*, 193b32–35). As will become more important in the next chapter, observe that Kepler placed limitations on God's omnipotence. God can only create that which is possible, although it is no slight to God's omnipotence to say that square circles and matterless quantity cannot be created. This limited notion of omnipotence became significant in the *Harmonice*, where Kepler argued that God could only create the best possible world, and this necessarily includes some imperfections.

18. Hence Kepler's conviction that astronomy must be founded on physical reasoning.

19. Stephenson (1987, 142) makes a similar point.

20. This is taken from Kepler's notes added to the *Mysterium* in 1621.

21. "Ad hanc obiectionem nihil respondere potui nisi hoc, rei cognatae similitudine, modum similem demonstrari, non plane rem eandem" (*KGW* XV 254). Unfortunately, Fabricius seemed uninterested in this answer, and so does not follow up on it.

22. Stephenson (1987, 110) makes a similar point.

Chapter 6
The *Harmonice mundi*

1. See Stephenson 1994, 90–97, for a discussion of Kepler's first theory of planetary harmonies.

2. For those interested in a more detailed exegesis, see Stephenson 1994, ch. 9.

3. This is problematic, for Ptolemy argued that the octave is slightly less than six tones. Ptolemy's octave is five tones plus two limmas, two sequences each with the ratio of 243:256 (Barker 1989b, 298–99). Ptolemy seemed content with this approximation: "The fact, then, that nature gave the circle of the signs of the zodiac a twelve-part constitution, is what we should rationally expect, since the complete *systēma* of a double octave is very close to twelve tones, and the interval of a tone was fitted to the twelfth part of the circle" (Barker 1989b, 383).

4. Ptolemy's titles read: (14) "Which are the first numbers in which the fixed notes of the complete *systēma* may be compared with the first spheres that there are in the universe" and (15) "How the ratios of the movements proper to each can be found in terms of numbers" (Barker 1989b, 362).

5. Here, indicating his full agreement, Kepler was quoting from Proclus.

6. This was Kepler's stated goal in the introduction to chapter 5, and an obvious lacuna left for him to fill after his discoveries in the *Astronomia*. He may not, however, have been confident in achieving this goal until late in the writing process, when he discovered his third law.

7. Kepler's harmonic archetypes also provide an argument against the Brahean system. Kepler wrote that unless we include the Earth's harmony, we lose the *genera* of the song (*Epitome*, 913; *KGW* VII, 314). He did not emphasize this claim, having promised a dying Brahe that he would complete the calculations for the Brahean system.

8. One wonders why the divine ear would not find these discrepancies excruciating.

9. Kepler described the *genera*, *mollis* and *durum*, in a manner that suggests that they correspond roughly to our major and minor modes. But Walker ([1967] 1978, 57–58) shows that Kepler used these terms in their traditional senses, where the *durum* contains B natural, and the *mollis* B flat.

10. In the rest of this section I follow Kepler's complex assignment of harmonies to planetary motions. A reader interested primarily in the philosophical aspects of Kepler's harmonies can skip ahead to the next section.

11. Wallace's translation of *proportio propria* in Kepler [1619] 1939, 1952.

12. The formula is simple algebra: the diverging movement is *Sa*:*Ip* (Superior aphelion to Inferior perihelion); the converging is *Sp*:*Ia*, and the private ratios are *Sa*:*Sp* and *Ia*:*Ip*, so the formula is simply

$$\frac{Sa}{Ip} \times \frac{Ia}{Sp} = \frac{Sa}{Sp} \times \frac{Ia}{Ip}.$$

Stephenson (1994) follows Kepler in referring to the multiplication and division of ratios as addition and subtraction respectively. Wallis translates these as "compound" and "inverse compound" respectively.

13. In proposition 8 Kepler assigned Saturn and Jupiter 1:2 and 1:3 because the ratio between the spheres of the cube is 1 to the root of 3, and the two nearest harmonies are 1:2 and 1:3. In proposition 9 Kepler used the formula to determine that the sum of Saturn's and Jupiter's private ratios is 2:3, which harmonically decomposes into 4:5 for the Saturn's private ratio, and 5:6 for Jupiter's (as a fifth decomposes into major and minor thirds).

14. Kepler obtained the ratios for Saturn by multiplying the square roots of the

ratios by 10. For Jupiter the square roots are multiplied by 1,000, Mars by 10. For Earth, the private ratio is multiplied by 3, then the squares are multiplied by 1,000. For Venus, multiply the private ratio by 40, and the square roots by 100. For Mercury, the private ratio is multiplied by 8, and the square roots by 10,000.

15. Wilson points out that Kepler should have corrected the Earth's eccentricity on the basis of astronomical considerations as well. Kepler's reduction of Brahe's calculation of horizontal solar parallax from 3′59″ to 1′ should have led to an alteration in the eccentricity (Wilson 1989, 251). For more on Horrocks, see Wilson 1989, 235–59.

16. "*And because we are seeking for all the mean motions measured in the same units, as all the proportions which have so far been confirmed between pairs of motions, and also all the planets' own motions, are set out in the common measure of their lowest common factor*" (*HM*, 484–85; *KGW* VI, 357; emphasis in original).

17. Stephenson (1994, 222) adds the prime factors of column 3 to show how it was calculated from columns 1 and 2.

18. In the *Harmonice*, tables 6.6 and 6.7 are displayed in one table.

19. Kepler displayed column 1 in table 6.8 in the same table on which tables 6.6 and 6.7 are based.

20. Let R = distance, P = period, AD = apparent diurnal arcs, TA = true diurnal arcs, and TD = true diurnal (linear) motion. We can use propositions 9 and 10 from Chapter 3.

Proposition 9: $$\frac{TD_1}{TD_2} \propto \left(\frac{TA_1}{TA_2} \times \frac{R_1}{R_2}\right).$$

Proposition 10: $$\frac{AD_1}{AD_2} \propto \left(\frac{TD_1}{TD_2} \times \frac{R_2}{R_1}\right).$$

Because a planet's true diurnal arcs are inversely proportional to the period, then by the harmonic law we get:

$$\left(\frac{R_1}{R_2}\right)^3 \propto \left(\frac{TA_2}{TA_1}\right)^2.$$

By proposition 9 we get:

$$\left(\frac{R_1}{R_2}\right)^3 \propto \left(\frac{TD_2}{TD_1} \times \frac{R_1}{R_2}\right)^2.$$

And by proposition 10:

$$\left(\frac{R_1}{R_2}\right)^3 \propto \left(\frac{AD_2}{AD_1}\right)^2.$$

Kepler distinguished between apparent and true diurnal arcs, stating, for example, that true diurnal motions can be determined by multiplying the distances by "the true (not the apparent) daily arcs of the eccentric" (*HM*, 412; *KGW* VI, 302). I have followed Kepler in speaking of the two as distinct, but as we can see by the preceding derivation, they are equivalent.

21. Kepler gave 1,878,483.

22. Kepler presented the data in tables 6.8 and 6.9 in one table.

23. Notice that this justifies Kepler's specific measure but not uniquely. Any measure that fulfills the criterion of order is similarly justified.

24. Perhaps too he was willing to admit an aspect of a theory that, while disturbing, is outweighed by the overall benefits of accepting the theory.

25. "[I]t is my pleasure to taunt mortal men with the candid acknowledgment that I am stealing the golden vessels of the Egyptians to build a tabernacle to my God" (*HM*, 391; *KGW* VI, 290).

Chapter 7
The *Epitome astronomiae Copernicana*: Kepler's Mature Physical Astronomy

1. Compare this to Kepler's saying that the *Harmonice* could "await its reader for a hundred years" (*HM*, 391; *KGW* VI, 290).

2. On the basis of his survey of seventeenth-century astronomers, Russell (1964, 1–24) maintains that the *Epitome* was Kepler's most influential theoretical work, in fact the most widely read book on astronomy in the European astronomical community.

3. Books I–III were written by 1615 and published in 1618. Kepler wrote book IV after writing the *Harmonice*, publishing the former in 1620. Books V–VII were published in 1621.

4. Books I–III address problems generated by the diurnal rotation of the Earth. Books IV–VI concern planetary and lunar theory, book VII the sphere of the stars.

5. See Koyré 1957, ch. 3, for an extended discussion of Kepler's arguments against the infinite universe. The example here is only one of several Kepler offered.

6. Here Kepler referred the reader to his *Hipparchus* (*KGW* XX.1), of which we only have fragments. See Swerdlow 1992, for a discussion of Kepler's *Hipparchus*, and a translation of a portion of it, the *Sciametria*.

7. Notice the difference between this and Kepler's discussion of planetary minds in the *Astronomia*. In the *Astronomia*, he was thinking physically, which is clear from his attempt to determine how a mind would have access to information necessary for maintaining the correct orbit.

8. Notice the warning against simply merging the disciplines, an attitude in keeping with Aristotle's *Posterior Analytics*.

9. See Stephenson 1987, 144–46, for a discussion of various misinterpretations of the third law.

10. As Voelkel (1994, 325) points out, Kepler was mistaken in believing the two theories to be equivalent. See Voelkel 1994, for a detailed analysis of the Kepler-Fabricius correspondence.

11. This problem is identified by Stephenson. For a much more detailed tour of Kepler's mathematical libration theory, and his derivation of the elliptical orbit, see Stephenson 1987, 147–57.

12. See my article (1999) for a more in-depth discussion of the significance of Kepler's balance analogy.

Conclusion
The Fate of Kepler's Philosophical Thought

1. Kepler met Bernegger in 1612, who later became Kepler's "best and most faithful friend" (Caspar [1959] 1993, 226).

2. ". . . habes infra in doctrina Eclipsium etiam ex hujus temporis observationibus documenta perspicua, motuum Solis, Lunae et Primi Mobilis non ad amussim Mathematicam aequalium, sed physicas minimas intensiones et remissiones recipientium extra ordinem" (*KGW* X, 44).

3. Jeremiah Horrock's distress over the perturbations and their implications for the status of astronomy are discussed by Wilson (1978, 248–55).

4. Stephenson similarly concludes Kepler's harmonies were ignored, with the notable exceptions of Jeremiah Horrocks and Giovanni Battista Riccioli: "Soon after birth they [Kepler's theories of planetary harmonies] fell into a kind of limbo, from which they have never emerged: too technical to be read by those who listen for the music of the spheres, and too peculiar to be taken seriously by scientists with the technical ability to understand them" (Stephenson 1994, 243).

5. I do not discuss the substantial influence of Kepler's astronomical achievements. On this subject, see Russell 1964 and Applebaum 1996.

6. I thank Jason Holt for pointing this out.

7. See Meli 1991.

Bibliography

Aiton, E. J. (1969). "Kepler's Second Law of Planetary Motion." *Isis* 60:75–90.

———. (1973). "Infinitesimals and the Area Law." In *Internationales Kepler-Symposium*, edited by F. Krafft, K. Meyer, and B. Stickler, pp. 285–305. Hildesheim: Gerstenberg.

Alexander, H. G., ed. (1956) *The Leibniz-Clarke Correspondence, Together with Extracts from Newton's "Principia" and "Optics."* Manchester: Manchester University Press.

Applebaum, W. (1996). "Keplerian Astronomy after Kepler: Researches and Problems." *History of Science* 34:451–504.

Bacon, F. (1887). *The Works of Francis Bacon*. Vol. 3. Edited by J. Spedding and R. E. Ellis. London: Longmans.

Barker, A. (1989a). *Greek Musical Writings*. Vol. 2, *Harmonic and Acoustic Theory*. New York: Cambridge University Press.

———, ed. (1989b). "Harmonics." In *Greek Musical Writings*, vol. 2, *Harmonic and Acoustic Theory*. pp. 275–391. New York: Cambridge University Press.

Barnes, J. ([1957, 1965] 1975). "Aristotle's Theory of Demonstration." In *Articles on Aristotle*, edited by J. Barnes, M. Scholfield, and R. Sorabji, pp. 65–87. London: Gerald Duckworth.

———. (1984). *The Complete Works of Aristotle: The Revised Oxford Translation*. Princeton: Princeton University Press.

Baumgardt, C. (1951). *Johannes Kepler: Life and Letters*. New York: Philosophical Library.

Beer, A. and P. Beer, eds. (1975). *Kepler: Four Hundred Years*. Vistas in Astronomy, vol. 18. New York: Pergamon Press.

Bennett, B. H. (1998). "Kepler's Answer to the Mystery: A New Cosmographical Epistemology." In *Mysterium Cosmographicum, 1596–1996*, edited by J. Folta, pp. 49–64. Acta Historiae Rerum Naturalium Necnon Technicarum: Prague Studies in the History of Science and Technology, vol. 2. Prague: National Technical Museum in Prague.

Boyd, R. ([1983] 1991). "On the Current Status of Scientific Realism." In *The Philosophy of Science*, edited by R. Boyd, P. Gasper, and J. D. Trout, pp. 195–222. Cambridge, Mass.: MIT Press. Reprinted from *Erkenntnis* 19:45–90.

Brahe, T. (1913–29). *Tychonis Brahe Dani Opera Omnia*. Edited by J.L.E. Dreyer. Hauniae: Libraria Gyldendaliana.

Buchdahl, G. (1972). "Methodological Aspects of Kepler's Theory of Refraction." *Studies in History and Philosophy of Science* 3:265–98.

Caspar, M. ([1959] 1993). *Kepler*. Translated by C. D. Hellman with a new introduction and references by O. Gingerich. New York: Dover.

Cassirer, E. ([1927] 1963). *The Individual and the Cosmos in Renaissance Philosophy*. New York: Harper & Row. Originally published as *Individuum und Kosmos in der Philosophie der Renaissance*. Studien der Bibliothek Warburg, vol. 10. Leipzig and Berlin: B. G. Teubner.

Clagett, M. ([1959] 1961). *The Science of Mechanics in the Middle Ages*. Madison: University of Wisconsin Press.

Cohen, I. B. ([1960] 1985). *The Birth of a New Physics*. New York: W. W. Norton.

———. (1972). "Newton and Keplerian Inertia: An Echo of Newton's Controversy with Leibniz." In *Science, Medicine and Society in the Renaissance: Essays to Honor Walter Pagel*, edited by A. G. Debus, pp. 199–211. New York: Science History Publications.

———(1975). "Kepler's Century: Prelude to Newton's." In *Kepler: Four Hundred Years*, edited by A. Beer and P. Beer, pp. 3–36. Vistas in Astronomy, vol. 18. New York: Pergamon Press.

Conger, G. P. ([1922] 1967). *Theories of Macrocosms and Microcosms in the History of Philosophy*. New York: Russell and Russell.

Copernicus, N. (1992). *On the Revolutions*. Translated by E. Rosen. Baltimore: John Hopkins University Press.

Cottingham, J. (1984). *Rationalism*. General Editor Justin Wintle. London: Paladin Books.

Crowe, Michael J. (1990). *Theories of the World from Antiquity to the Copernican Revolution*. New York: Dover.

de Cusa, N. (1932–). *Nicolai de Cusa Opera Omnia/iussu et auctoritate Academiae Litterarum Heidelbergensis ad codicum fidem edita*. Vol. 1, *De docta ignoranta*. Edited by Ernst Hoffman and Raymond Klibansky. Lipsiae: in aedibus Felicis Meiner.

Descartes, R. (1960). *Meditation on First Philosophy*. Translated by L. F. Lafteur. Indianapolis: Bobbs-Merrill.

Drake, S. (1975). "Kepler and Galileo." In *Kepler: Four Hundred Years*, edited by A. Beer and P. Beer, pp. 237–47. Vistas in Astronomy, vol. 18. New York: Pergamon Press.

Drake, S. and Drabkin, I. E. (1969). *Mechanics in Sixteenth Century Italy: Selections from Tartaglia, Benedetti, Guido Ubaldo, and Galileo*. Madison: University of Wisconsin Press.

Dreyer, J.L.E. (1890). *Tycho Brahe, A Picture of Scientific Life and Work in the Sixteenth Century*. Edinburgh: Adam and Charles Black.

———([1906] 1953). *A History of Astronomy from Thales to Kepler*. New York: Dover. Reprinted from *History of the Planetary Systems from Thales to Kepler*. Cambridge: Cambridge University Press.

Duhem, P. ([1908] 1969). *To Save the Phenomena: An Essay on the Idea of Physical Theory from Plato to Galileo*. Translated by E. Doland and C. Maschler. Chicago: University of Chicago Press. Originally published as "*ΣΩΖΕΙΝ ΤΑ ΦΑΙΝΟΜΕΝΑ*: Essai sur la notion de théorie physique de Platon à Galilée." *Annales de philosophie chrétienne*, 79/156 (ser. 4, VI): 113–38, 277–302, 352–77, 482–514, 576–92.

Field, J. V. (1988). *Kepler's Geometrical Cosmology*. Chicago: University of Chicago Press.

Gardner, M. R. (1983). "Realism and Instrumentalism in Pre-Newtonian Astronomy." In *Testing Scientific Theories*, edited by John Earman, pp. 207–65. Minnesota Studies in the Philosophy of Science, vol. 10. Minneapolis: University of Minnesota.

Gilbert, N. W. (1960). *Renaissance Concepts of Method*. New York: Columbia University Press.

Gingerich, O. ([1964] 1993). "The Computer versus Kepler." *American Scientist*, 52 (2):218–16. Reprinted in *The Eye of Heaven: Ptolemy, Copernicus, Kepler*, pp. 357–66. New York: American Institute of Physics.

———. (1973). "Kepler's Treatment of Redundant Observations or, the Computer versus Kepler Revisted." In *Internationales Kepler-Symposium*, edited by F. Krafft, K. Meyer and B. Stickler, pp. 307–14. Hildesheim: Gerstenberg.

——— (1975a). "The Origins of Kepler's Third Law." In *Kepler: Four Hundred Years*, edited by A. Beer and P. Beer, pp. 595–601. *Vistas in Astronomy*, vol. 18. New York: Pergamon Press.

———. (1975b). "Kepler's Place in Astronomy." In *Kepler: Four Hundred Years*, edited by A. Beer and P. Beer, pp. 261–78. *Vistas in Astronomy*, vol. 18. New York: Pergamon Press.

———. (1992). "Kepler, Galilei, and the Harmony of the World." In *Music and Science in the Age of Galileo*, edited by V. Coelho, pp. 45–63. Netherlands: Kluwer.

Gingerich, O., and J. R. Voelkel. (1998). "Tycho Brahe's Copernican Campaign." *Journal for the History of Astronomy* 29:1–34.

Gingerich, O., and R. Westman, (1988). *The Wittich Connection: Conflict and Priority in Late Sixteenth-Century Cosmology*. Transactions of the American Philosophical Society, 78, pt. 7. Philadelphia.

Grafton, A. (1973). "Michael Maestlin's Account of Copernican Planetary Theory." *Proceedings of the American Philosophical Society* 117 (6):523–50.

Hallyn, F. ([1987] 1990). *The Poetic Structure of the World*. Translated by Donald M. Leslie. New York: Zone Books. Originally published as *La structure poétique du mond: Copernic, Kepler*. Paris: Éditions du Seuil.

Hamilton, E, and H. Cairns, eds. (1961). *The Collected Dialogues of Plato*. Princeton: Princeton University Press.

Harper, W. , and G. Smith. (1995). "Newton's New Way of Inquiry." In *The Creation of Ideas in Physics*, edited by J. Leplin, pp. 113–66. Netherlands: Kluwer.

Henry, J. (1997). *The Scientific Revolution and the Origins of Modern Science*. New York: St. Martin's Press.

Holton, G. (1956). "Johannes Kepler's Universe: Its Physics and Metaphysics." *American Journal of Physics* 24:340–51.

Jardine, N. (1979). "The Forging of Modern Realism: Clavius and Kepler against the Sceptics." *Studies in History and Philosophy of Science* 10:141–73.

———. (1984). *The Birth of History and Philosophy of Science: Kepler's "A Defense of Tycho against Ursus" with Essays on Its Provenance and Significance*. Cambridge: Cambridge University Press.

Jarrell, R. A. (1971). "The Life and Scientific Work of the Tübingen Astronomer Michael Maestlin, 1550–1631." Ph.d. diss., University of Toronto.

Kenny, A., trans. and ed. (1970). *Descartes' Philosophical Letters*. Oxford: Oxford University Press.

Kepler, J. ([1596] 1981). *The Secret of the Universe*. Translated by A. M. Duncan with notes by E. J. Aiton. New York: Abaris Books.

———. ([1609] 1992). *New Astronomy*. Translated by W. H. Donahue. New York: Cambridge University Press.

Kepler, J. ([1618–21] 1952). *Epitome of Copernican Astronomy: Books IV and V.* Translated by C. G. Wallis. In *Great Books of the Western World*, vol. 16, edited by R. M. Hutchins, pps. 839–1004. Chicago: Encyclopaedia Britannica.

———. ([1619] 1939, 1952). *The Harmonies of the World: V.* Translated by C. G. Wallis. In *Great Books of the Western World*, vol. 16, edited by R. M. Hutchins, pps. 1005–85. Chicago: Encyclopaedia Britannica.

———. ([1619] 1997). *The Harmony of the World.* Translated with an introduction and notes by E. J. Aiton, A. M. Duncan, and J. V. Field. Philadelphia: American Philosophical Society.

———. (1858–91). *Joannis Kepleri astronomi opera omnia.* Edited by C. Frisch. Frankfurt: Heyder and Zimmer.

———. ([1858] 1984). *Apologia pro Tychone contra Ursum.* Translated by N. Jardine. In *The Birth of History and Philosophy of Science: Kepler's "A Defense of Tycho against Ursus" with Essays on its Provenance and Significance*, pp. 133–202. Cambridge: Cambridge University Press.

———. (1937–). *Johannes Kepler Gesammelte Werke.* Edited by W. V. Dyck, M. Caspar, and F. Hammer. Munich: Beck.

Koestler, A. ([1959] 1968). *The Sleepwalkers.* London: Hutchinson.

Koyré, A. (1957). *From the Closed World to the Infinite Universe.* Baltimore: Johns Hopkins Press.

———. ([1961] 1973). *The Astronomical Revolution: Copernicus—Kepler—Borelli.* Paris: Hermann. Originally titled *La Révolution astronomique.*

Kozhamthadam, J. (1994). *The Discovery of Kepler's Laws: The Interaction of Science, Philosophy, and Religion.* Notre Dame, Ind.: University of Notre Dame Press.

Kristeller, P. O. ([1943] 1964). *The Philosophy of Marsilio Ficino.* New York: Columbia University Press.

———. (1979). *Renaissance Thought and Its Sources.* New York: Columbia University Press.

Kuhn, T. S. (1957). *The Copernican Revolution: Planetary Astronomy in the Development of Western Thought.* Cambridge, Mass.: Harvard University Press.

Lakatos, I., and Zahar, E. ([1976] 1978). "Why Did Copernicus's Programme Supersede Ptolemy's?" In *The Methodology of Scientific Research Programmes: Philosophical Papers*, vol. 1, edited by J. Worrall and G. Curry, pp. 168–92. Cambridge: Cambridge University Press. Reprinted from *The Copernican Achievement*, edited by R. Westman, pp. 354–83. Los Angeles: University of California Press.

Lattis, J. M. (1994). *Between Copernicus and Galileo: Christoph Clavius and the Collapse of Ptolemaic Cosmology.* Chicago: University of Chicago Press.

Leibniz, G. W. (1962). *G. W. Leibniz Mathematische Schriften, Vol. VI.* Edited by C. I. Gerhardt. Hildesheim: Georg Olms Verlagsbuchhandlung.

Lindberg, D.C. (1986). "The Genesis of Kepler's Theory of Light: Light Metaphysics from Plotinus to Kepler." *Osiris*, 2d ser., 2:5–42.

Martens, R. (1998). "Kepler's Use of Archetypes in His Defense against Aristotelian Scepticism." In *Mysterium Cosmographicum 1596–1996*, edited by J. Folta, pp. 74–89. Acta Historiae Rerum Naturalium Necnon Technicarum: Prague

Studies in the History of Science and Technology, vol. 2. Prague: National Technical Museum in Prague.

———. (1999). "Kepler's Solution to the Problem of a Realist Celestial Mechanics." *Studies in History and Philosophy of Science* 30 (3):377–94.

Meli, D. B. (1991). "Public Claims, Private Worries: Newton's Principia and Leibniz's Theory of Planetary Motion." *Studies in History and Philosophy of Science* 22 (3):415–49.

Methuen, C. (1998). *Kepler's Tübingen: Stimulus to a Theological Mathematics*. Hampshire: Ashgate.

Mittelstrass, J. (1972). "Methodological Elements of Keplerian Astronomy." *Studies in History and Philosophy of Science* 3:203–32.

Neugebauer, O. ([1957] 1969). *The Exact Sciences in Antiquity*. 2d ed. New York: Dover.

Pauli, W. (1955). "The Influence of Archetypal Ideas on the Scientific Theories of Kepler." In *The Interpretation of Nature and the Psyche*, pp. 147–240. New York: Pantheon.

Rose, P. L., and S. Drake. (1971). "The Pseudo-Aristotelian Questions of Mechanics in Renaissance Culture." *Studies in the Renaissance* 18:65–104.

Rosen, E. (1984). "Kepler's Attitude toward Astrology and Mysticism." In *Occult and Scientific Mentalities in the Renaissance*, edited by B. Vickers, pp. 253–72. New York: Cambridge University Press.

———. (1986). *Three Imperial Mathematicians: Kepler Trapped between Tycho Brahe and Ursus*. New York: Abaris.

Russell, J. L. (1964). "Kepler's Laws of Planetary Motion: 1609–1666." *British Journal for the History of Science* 2:1–24.

Schiaparelli, G. (1877). *Die homocentrischen Sphären des Eudoxus, des Kallippus und des Aristoteles*. Translated by W. Horn. Leipzig: B. G. Teubner.

Schmitt, C. B. (1975). "Philosophy and Science in Sixteenth-Century Universities: Some Preliminary Comments." In *The Cultural Context of Medieval Learning*, edited by J. E. Murdoch and E. D. Sylla, pp. 485–537. Dordecht: D. Reidel.

———. (1983). *Aristotle and the Renaissance*. Cambridge, Mass.: Harvard University Press.

Shapin, S. (1996). *The Scientific Revolution*. Chicago: University of Chicago Press.

Simon, G. (1975). "Kepler's Astrology: The Direction of a Reform." In *Kepler: Four Hundred Years*, edited by A. Beer and P. Beer, pp. 439–48. Vistas in Astronomy, vol. 18, New York: Pergamon Press.

Small, R. ([1804] 1963). *An Account of the Astronomical Discoveries of Kepler*. Madison: University of Wisconsin Press.

Smart, J.J.C. (1968). *Between Science and Philosophy*. New York: Random House.

Stephenson, B. (1987). *Kepler's Physical Astronomy*. New York: Springer-Verlag. Reprinted by Princeton University Press, 1994.

———. (1994). *The Music of the Heavens: Kepler's Harmonic Astronomy*. Princeton: Princeton University Press.

Straker, S. (1981). "Kepler, Tycho, and the 'Optical Part of Astronomy': The Genesis of Kepler's Theory of Pinhole Images." *Archive for History of Exact Sciences* 24:267–93.

Strong, E. (1966). *Procedures and Metaphysics*. Hildesheim: Georg Olms.

Swerdlow, N. M. (1973). "The Derivation and First Draft of Copernucus's Planetary Theory: A Translation of the *Commentariolus* with Commentary." *Proceedings of the American Philosophical Society* 117:423–512.

Swerdlow, N. M., and Neugebauer, O. (1984). *Mathematical Astronomy in Copernicus's "De revolutionibus."* 2 vols. New York: Springer-Verlag.

———. (1992). "Shadow Measurement: The *Sciametria* from Kepler's *Hipparchus*—a Translation with Commentary." In *An Investigation of Difficult Things: Essays on Newton and the History of the Exact Sciences*, edited by P. M. Harman and Alan E. Shaprio, 19–70. Cambridge: Cambridge University Press.

Tannery, P. (1932–88). *Marin Mersenne Correspondance*. Published by Mme. Paul Tannery, edited and annotated by Cornelis de Waard, with René Pintard. Paris: Beau-Chesne.

Taub. L. C. (1993). *Ptolemy's Universe*. Chicago: Open Court.

Thoren, V. E. (1990). *The Lord of Uraniborg: A Biography of Tycho Brahe*. Cambridge: Cambridge University Press.

Toomer, G. J., trans. (1984). *Ptolemy's Almagest*. New York: Springer-Verlag.

Voelkel, J. R. (1994). *The Development and Reception of Kepler's Physical Astronomy, 1593–1609*. Ann Arbor, Mich.: UMI.

Wahba, M., and M. Abousenna. (1966). *Averroes and the Enlightenment*. New York: Prometheus Books.

Walker, D. P. ([1967] 1978). "Kepler's Celestial Music." In *Studies in Musical Science in the Late Renaissance*, pp. 34–62. London: Warburg Institute.

Wallace, W. A. (1981). *Prelude to Galileo: Essays on Medieval and Sixteenth-Century Sources of Galileo's Thought*. Boston: D. Reidel.

———. (1988). "Traditional Natural Philosophy." In *The Cambridge History of Renaissance Philosophy*, edited by C. B. Schmitt, Q. Skinner, E. Kessler, and J. Kraye, pp. 201–35. New York: Cambridge University Press.

Westman, R. S. (1972). "Kepler's Theory of Hypothesis and the 'Realist Dilemma.'" *Studies in History and Philosophy of Science* 3:233–264.

———. (1975). "Three Responses to the Copernican Theory: Johannes Praetorius, Tycho Brahe, and Michael Maestlin." In *The Copernican Achievement*, edited by Robert S. Westman, pp. 285–345. Berkeley: University of California Press.

———. (1980). "The Astronomer's Role in the Sixteenth Century: A Preliminary Survey." *History of Science* 18:105–47.

———. (1984). "Nature, Art, and Psyche: Jung, Pauli, and the Kepler-Fludd Polemic." In *Occult and Scientific Mentalities in the Renaissance*, edited by B. Vickers, pp. 177–229. New York: Cambridge University Press.

Whewell, W. (1968). *William Whewell's Theory of Scientific Method*. Edited by R. E. Butts. Pittsburgh: University of Pittsburgh Press.

Whiteside, D. T. (1974). "Keplerian Planetary Eggs, Laid and Unlaid, 1600–1605." *Journal for the History of Astronomy* 5:1–21.

Wilson, C. (1968). "Kepler's Derivation of the Elliptical Path." *Isis* 59:5–25.

———. (1970). "From Kepler's Laws, So-Called, to Universal Gravitation: Empirical Factors." *Archive for History of Exact Science* 6:115–16.

——— (1972). "The Inner Planets and the Keplerian Revolution." *Centaurus* 17:205–48.

———. (1974). "Newton and Some Philosophers on Kepler's "Laws." *Journal of the History of Ideas* 35:231–58.

———. (1978). "Horrocks, Harmonies, and the Exactitude of Kepler's Third Law." In *Science and History. Studia Copernicana*, 16: 235–59.

———. (1989). *Astronomy from Kepler to Newton*. London: Variorum.

Yavetz, I. (1998). "On the Homocentric Spheres of Eudoxus." *Archive for History of Exact Sciences* 52 (3):221–78.

Index